Springer Handbook of
Auditory Research

Series Editors: Richard R. Fay and Arthur N. Popper

Richard R. Fay
Arthur N. Popper

Editors

Comparative Hearing: Mammals

With 83 illustrations

Springer-Verlag
New York Berlin Heidelberg London Paris
Tokyo Hong Kong Barcelona Budapest

Richard R. Fay
Parmly Hearing Institute and
Department of Psychology
Loyola University of Chicago
Chicago, IL 60626 USA

Arthur N. Popper
Department of Zoology
University of Maryland
College Park, MD 20742 USA

Series Editors: Richard R. Fay and Arthur N. Popper

Cover illustration: The organ of Corti from the gerbil (top) and mole rat. See Figure 5.7.

Library of Congress Cataloging-in-Publication Data
Comparative hearing: mammals / Richard R. Fay, Arthur N. Popper,
    editors.
        p.   cm.—(Springer handbook of auditory research)
        Includes bibliographical references and index.
        ISBN 0-387-97841-0 (acid-free paper: New York).—ISBN
3-540-97841-0 (acid-free paper: Berlin)
        1. Hearing.   2. Physiology, Comparative.   3. Mammals—Physiology.
    I. Fay, Richard R.   II. Popper, Arthur N.   III. Series.
    QP461.C64   1994
    599′.01825—dc20   93-43309

Printed on acid-free paper.

Production managed by Henry Krell; manufacturing supervised by Gail Simon.
Typeset by Asco Trade Typesetting Ltd., Hong Kong.
Printed and bound by Edwards Brothers, Inc., Ann Arbor, MI.
Printed in the United States of America.

9 8 7 6 5 4 3 2 1

ISBN 0-387-97841-0 Springer-Verlag New York Berlin Heidelberg
ISBN 3-540-97841-0 Springer-Verlag Berlin Heidelberg New York

Series Preface

The *Springer Handbook of Auditory Research* presents a series of comprehensive and synthetic reviews of the fundamental topics in modern auditory research. The volumes are aimed at all individuals with interests in hearing research including advanced graduate students, postdoctoral researchers, and clinical investigators. The volumes are intended to introduce new investigators to important aspects of hearing science and to help established investigators to better understand the fundamental theories and data in fields of hearing that they may not normally follow closely.

Each volume is intended to present a particular topic comprehensively, and each chapter will serve as a synthetic overview and guide to the literature. As such, the chapters present neither exhaustive data reviews nor original research that has not yet appeared in peer-reviewed journals. The volumes focus on topics that have developed a solid data and conceptual foundation rather than on those for which a literature is only beginning to develop. New research areas will be covered on a timely basis in the series as they begin to mature.

Each volume in the series consists of five to eight substantial chapters on a particular topic. In some cases, the topics will be ones of traditional interest for which there is a substantial body of data and theory, such as auditory neuroanatomy (Vol. 1) and neurophysiology (Vol. 2). Other volumes in the series will deal with topics which have begun to mature more recently, such as development, plasticity, and computational models of neural processing. In many cases, the series editors will be joined by a co-editor having special expertise in the topic of the volume.

Richard R. Fay
Arthur N. Popper

Preface

Hearing researchers often use species other than humans as the subject of experimentation. One major goal of much of this research is to develop and evaluate animal models for both normal and pathological aspects of human hearing. Another goal of much of this animal research is not the development of animal models, but a deeper understanding of hearing as a general biological phenomenon. This approach, termed "comparative hearing research," investigates the structures, physiological functions, and hearing abilities of various species in order to determine the fundamental principles by which structures determine functions, and to help clarify the evolutionary history of hearing among animals. Comparative hearing research establishes the context within which animal models can be developed, evaluated, validated, and successfully applied, and is therefore of fundamental importance to hearing research in general. Much of the research discussed in this volume is motivated by these concerns.

The chapters of this volume review what is understood about the similarities and differences among mammals in their sense of hearing, defined behaviorally, and the structures of the ear that condition and transform sound wave-forms before they are transduced within the inner ear. Later volumes in this series will focus on comparative auditory research in invertebrates, fishes, amphibians, reptiles, and birds.

Fay (Chapter 1) introduces some of the fundamental ideas that motivate and inform comparative hearing research, and places each of the other chapters in this general context. Long (Chapter 2) reviews and discusses the fundamental hearing capabilities of mammals as determined by psychophysical methods, emphasizing the experimental designs used to investigate the relationships between human and nonhuman hearing. Brown (Chapter 3) presents the principles by which mammals locate sound sources in three-dimensional space, with special treatments of the sound source, receiver, and environmental characteristics most important for the accurate determination of source azimuth, elevation, and distance. Stebbins and Moody (Chapter 4) focus on the perception of complex, vocalization sounds in nonhuman primates using both traditional psychoacoustic paradigms and those designed to investigate higher order, or cognitive, functions in auditory perception. Echteler, Fay, and Popper (Chapter 5) present the major structural and mechanical features of the cochlea that play important roles in sound transduction and transformation, and discuss how these features vary among mammals. Finally, Rosowski (Chapter 6) synthesizes the mechanical principles that determine the frequency-dependent characteristics of the outer and middle ears as they transmit acoustic power to the inner ear.

The goal of each chapter is to present what is known about the diversity of hearing functions among mammals so that we may better understand the relationships between the structures of ears and the hearing abilities of the animals most closely related to humans and most often used as models for hearing in humans. One of the themes of this volume concerns the ways in which we study and conceptualize the differences among mammals, and the similarities and differences between nonhuman mammals and humans.

Richard R. Fay
Arthur N. Popper

Contents

Contributors

Charles H. Brown
Department of Psychology, University of South Alabama, Mobile, AL 36688
USA

Stephen M. Echteler
Auditory Physiology Laboratory, Northwestern University, Evanston, IL
60208 USA

Richard R. Fay
Department of Psychology and Parmly Hearing Institute, Loyola University
of Chicago, Chicago, IL 60626 USA

Glenis R. Long
Department of Audiology and Speech Sciences, Purdue University, West
Lafayette, IN 47907 USA

David B. Moody
Department of Psychology and Kresge Hearing Research Institute, University of Michigan, Ann Arbor, MI 48109 USA

Arthur N. Popper
Department of Zoology, University of Maryland, College Park, MD 20742
USA

John J. Rosowski
Eaton-Peabody Laboratory and Department of Otolaryngology, Massachusetts Eye and Ear Infirmary, Boston, MA 02114 and The Research Laboratory of Electronics, Massachusetts Institute of Technology, Cambridge, MA
02138 USA

William C. Stebbins
Department of Psychology and Kresge Hearing Research Institute, University of Michigan, Ann Arbor, MI 48109 USA

1
Comparative Auditory Research

RICHARD R. FAY

1. Introduction

The chapters of this volume review what is understood about the similarities and differences among mammals in their sense of hearing, defined behaviorally, and the structures of the ear that condition and transform sound wave-forms before they are transduced into patterns of neural activity.

Long (Chapter 2) reviews the present knowledge about the fundamental hearing capabilities of mammals as determined by psychophysical methods, with a particular emphasis on the experimental designs used to investigate the relationships between human and nonhuman hearing. Brown (Chapter 3) discusses the cues and principles by which mammals locate sound sources, with an emphasis on the sound source, receiver, and environmental characteristics most important for the accurate determination of source azimuth, elevation, and distance. Stebbins and Moody (Chapter 4) focus on the present understanding of hearing and the perception of complex, vocalization sounds in nonhuman primates using both traditional paradigms and those designed to investigate higher order, or cognitive, functions in auditory perception. Echteler, Fay and Popper (Chapter 5) present the major structural and mechanical features of the cochlea that play important roles in the transformation of stapes movement to the adequate stimuli impinging on the hair cell receptor array, and discuss how these features vary among mammals. Rosowski (Chapter 6) synthesizes the mechanical principles that determine the frequency-dependent characteristics of the outer and middle ears as they transmit acoustic power to the inner ear. The goal of each chapter is to present what is known about the diversity of hearing among mammals so that we may better understand the relationships between the structures of ears and the hearing abilities of the animals most closely related to humans and most often used as models for hearing in humans. Reviews of mammalian auditory neurophysiology and neuroanatomy can be found in Popper and Fay (1992) and Webster, Popper, and Fay (1992), respectively.

This chapter outlines a general context within which observations on the hearing and auditory systems of diverse mammalian species, including humans, can be better understood. One of the themes of this chapter and this volume concerns the ways in which we study and conceptualize the differences among mammals, and the similarities and differences between non-human mammals and humans.

1.1 Comparative Hearing Research

Hearing research often makes use of species other than humans as the subject of experimentation. In animal studies, the assumption usually is that the experiment aims to provide data that will be compared with that of another species, either humans or some other species. If the major goal of the research is to compare results with humans, the species is used as an "animal model," and the value of the results will depend upon how closely they resemble those expected from humans. In some cases, comparable data may be obtainable from the model species and humans, and the value of the model can be directly evaluated. Scalp-recorded averaged evoked response studies are examples (see Kraus and McGee 1992). In many cases, however, there is no clear indication of the validity of the model. For example, our understanding of the ways in which sound features are represented in patterns of neural activity in humans arises from studies on single cells in many species of mammals and nonmammals. Studies of this kind are not carried out on human subjects, and in order to apply these measurements to questions of human hearing, we have had simply to assume similarities between these responses and those occurring in humans.

Some animal research has as its goal not the development of an animal model for some aspect of human hearing, but an increased understanding of hearing as a general biological phenomenon. This sort of comparative research investigates the diverse structures, physiological functions, and hearing abilities of various species in order to determine the fundamental principles by which structures determine functions and to help clarify the evolutionary history of hearing among animals. This research program is termed "comparative hearing research," to distinguish it from the "animal model" approach. Much of the research discussed in this volume is motivated by these concerns (see Webster, Fay, and Popper 1992).

A major goal of the study of hearing is to explain how the human auditory system normally functions and to identify the causes of and treatments for hearing impairment. Experimental approaches to these questions require the development and application of animal models, and their validity and reliability will determine the success of this effort. Comparative hearing research establishes the context within which animal models can be developed, evaluated, validated, and successfully applied and is, therefore, of fundamental importance to hearing research in general. For example, the importance of the observation that hair cells may regenerate in the bird cochlea cannot be evaluated for its potential impact on human hearing without a comparative and evolutionary context within which observations on both birds and mammals can fit and be fully understood. Similarly, observations on the effects of ototoxic drugs on hair cells and on hearing sensitivity of the chinchilla can only be of value to questions of human hearing when we know how to generalize observations such as these from the chinchilla to the human auditory system. We must have confidence, for example, that the same or similar biochemical and biophysical systems operate in the chinchilla and human cochleae, that localized cochlear damage has similar effects on sound perception in the chinchilla and in humans, and that we are able to measure hearing processes in chinchillas in the same quantitative terms used for humans. Only then will we be able to use fully the chinchilla data to help understand human hearing and its disorders. This confidence arises from the general understanding of structure/functional relationships that has come from comparative hearing research.

1.2 The Sense of Hearing

1.2.1 Definitions

Hearing is defined as the act of perceiving sound, and a sense of hearing can only be demonstrated and described through an analysis of sound's effects on behavior (see Webster 1992). For these reasons, the most important questions in hearing research arise from behavioral observations. Psychophysics is a field of experimental psychology that investigates the relationships between behavioral responses indicating a judgment or decision and the physical dimensions of stimuli upon which these decisions are based (see Green and Swets 1966). Animal psychoacoustics is the study of hearing in nonhuman animals, using psychophysical methods in combination with conditioning paradigms (see Fay 1988 for a review). Ideally, the primary difference between animal and human psychophysics is simply that animals must be trained and motivated to make the responses that cooperative human observers readily give after instruction. Over 100 years of research on animal conditioning and learning has developed a powerful technology for behavior analysis and control, and hundreds of animal psychoacoustical studies have been carried out using this technology (reviewed in Fay 1988; see Long, Chapter 2; Brown, Chapter 3; Stebbins and Moody, Chapter 4).

1.2.2 The Uses of Comparative Psychoacoustics

Why do we do comparative psychoacoustic research? Why cannot neurophysiological experiments substitute for animal psychoacoustics? Since hearing is defined as a behavior, it must be studied using behavioral methods such as those of psychophysics. Although we can define auditory sensitivity and bandwidth using physiological measure in humans and other animals (e.g., auditory evoked potentials; see Kraus and McGee 1992), our definition of the sense of hearing requires behavioral measurements. Human hearing has been defined by the results of numerous psychophysical studies carried out for over 100 years (see Yost, Popper, and Fay 1993). In order to use any animal as a model in hearing research, we must define its sense of hearing similarly.

Animal psychoacoustics is also valuable in two other ways: it provides logical links between animal neurophysiology and human hearing, and it defines a biological context within which human psychophysical data gain additional meaning and importance. Our understanding of human auditory neurophysiology comes primarily from experiments on single auditory neurons of nonhuman species (see Popper and Fay 1992). As comparative neurophysiological data accumulate to establish some of the general principles that apply to all or most vertebrates, differences among species may be described and understood in terms of these principles. For example, auditory nerve fibers for all vertebrates investigated show phase locking to low-frequency tones (see Ruggero 1992). Species can be compared and characterized by quantitative deviations, or the lack of them, from the usual patterns. With this understanding, we can develop an estimation or prediction of the response of auditory neurons in humans. These estimations then become assumptions that we make on the basis of comparative research. Are there ways to test these assumptions?

Aside from conducting experiments on the acoustic response properties of human auditory neurons (experiments that are not likely to be done), we can, at least tentatively, investigate human neurophysiology through psychophysical experiments. For example, psychophysical tuning curves (PTC; Serafin, Moody, and Stebbins 1982) are the result of psychophysical masking experiments designed to reveal the frequency selectivity of hypothetical filters used in sound detection (see Moore 1993; Long, Chapter 2; Stebbins and Moody, Chapter 4). It has been observed that the shapes and other characteristics of human PTCs resemble neurophysiological tuning curves of other mammals. This resemblance has led to the conclusion that human PTCs reveal characteristics of human neurophysiological tuning curves. (However, see Long, Chapter 2 for the difficulties of interpreting the shapes of PTCs.) This conclusion may be of limited value since it is based on the assumption that physiological tuning curves for the monkey are similar to those of humans. If this is true, why bother to measure psychophysical tuning curves at all?

What we would like to be able to show is that, in general, properly designed behavioral experiments on humans may directly reveal their neurophysiological causes. A direct test of this idea requires, for example, that we measure PTCs and neurophysiological tuning curves in the same, nonhuman species using comparable methods. These results would directly demonstrate the relationship between psychophysical and neurophysiological data, and this relationship would then become a hypothesis for the relationship between neural and psychophysical data in humans. This hypothesis could be modified and refined as these experiments are carried out in more species. If the relationships between physiology and psychophysics are entirely species specific, we would have to conclude that there is little hope of using comparative data to study human hearing. However, if the relationships show little variation across species, or variation that can be accounted for by ear structure or other factors that are observable in humans, we will be able to estimate them in humans with some confidence. Thus, human neurophysiology can be estimated from psychophysics once the relationships between neurophysiological and psychophysical data obtained for the same species are understood in general. In this sense, comparative physiological and psychoacoustic studies are necessary for a full understanding of human hearing.

A second way in which animal psychoacoustics is valuable is in enhancing the importance and meaning of human psychoacoustic data. If there was no chance of obtaining psychophysical data on nonhumans, human psychophysics would be a somewhat isolated field that simply describes the relationships between human performance and acoustic variables. The practical reality of animal psychoacoustics creates a wider, biological arena within which human psychoacoustic data can also be used for comparison with other species and as a way to draw conclusions about their physiological causes. In short, animal psychoacoustics helps place human psychoacoustic data within the realms of evolutionary biology and neuroscience. This being the case, it is possibly surprising that research in animal psychoacoustics is supported in only a handful of laboratories worldwide. One of its difficulties is that it is easy to argue for its value in general, but is sometimes more difficult to do so in specific cases. In any case, animal psychoacoustics helps create a biological context within which data on human hearing can be better understood and more widely applied.

1.2.3 What Do Animals Listen To?

It probably seems quite clear to most of us what we listen to all day: speech, music, the sounds of other animals communicating, and some nonbiological environmental sounds. If we ask what most animals or any given species listen to (Myrberg 1981), we are likely to decide that they listen to intraspecific communication sounds (see Brown, Chapter 3; Stebbins and Moody, Chapter 4). With this judgment comes the assumption that the ear and auditory system appear to be elements in an acoustic communication system and have evolved to receive and decode the messages and meanings of communication sounds such as speech and other vocalizations. While this view is valuable for studying acoustic communication per se, it can also divert our attention from perhaps more fundamental functions of hearing that are shared by all mammals, and perhaps all animals with a sense of hearing. What are these functions?

The question "What do animals look at?" is not often asked. We seem ready to accept that vertebrate visual systems have the general function of informing the organism about the objects in the environment that produce and scatter visible light. It would seem that animals "look at" everything visible in their immediate world and probably form an image in memory which represents these objects and their relationships with one another. Although many mammal species communicate by visual cues, we do not view the visual system primarily as a communication system and do not attempt to explain structures of the eye or details of ganglion cell physiology, for example, as adaptations for processing certain visual communication signals. Perhaps we should view hearing as similar to vision in this respect: its most general, primitive function is to obtain information on the identities and locations of the objects in the immediate world that produce or scatter sound. In other words, the most primitive function of hearing is to inform the organism about objects and events in the environment so that it can draw the right conclusions about them and act appropriately. It would seem that the organisms best able to draw the right conclusions about objects in the immediate environment are more fit than those that fail to perceive reality and would be expected to live longer and reproduce more successfully. (For those uncomfortable with the notion of "reality" as used above, it is defined here as whatever is confirmable by further exploration using hearing and the other senses.)

Some sound sources are relatively continuous and chaotic (e.g., sounds of the wind, rain, and flowing water) while others may be brief or spectrally and temporally patterned or have communicative value. Many of these sounds may occur simultaneously and reach the ears as complex mixtures. The problem facing the ears and brain is to segregate the acoustic components from the several sources into groups that belong to the appropriate source. Viewing the problem in this way, we could say that all sources that may produce or scatter audible sound are equally "biologically significant" in the sense that no source can be identified or localized without significant processing of the simultaneous sounds from the other sources. In an analogy with vision, the "ground" must be as well processed as the "figure" in order for the figure to be perceived as an object, and in many cases, the structure of the "ground" helps define the "figure." In other words, both "signals" and "noise" require analysis, because the sources of both must be understood to some degree before the source of either one can be identified and located in the presence of the other. The answer to the question. "What do animals listen to?" is perhaps "all sounds," including what is termed "ambient noise." It is suggested here that the fundamental function of hearing is to characterize the acoustic ambience and its perturbations created by all sources and scatterers (e.g., Rogers et. al. 1989).

As important as the coded information that one organism communicates to another is the ability of each to perceive the other as an entity, object, or source that is maintained in perception as an object during the passage of time and during relative movements between the perceiver and the perceived. It makes sense that once such a perceptual system had evolved, it would also be recruited for communication purposes. It could be argued, however, that a more complete understanding of hearing as a biological phenomenon requires that we recognize its most primitive, or shared, role: the identification and localization of sound sources within the "auditory scene" (Bregman 1990).

The physical characteristics of sound sources clearly have not changed during vertebrate evolution. This means that we should expect to find auditory mechanisms in all species that accurately encode the physical features of sound sources so that perceptual objects can be formed, corresponding to those formed through other senses, and confirmed through behavioral interaction with the environment. Accurate encoding of acoustic features depends on the structures and physiological processes of the auditory periphery. Thus, this volume presents reviews and discussions of the acoustic structures of natural environments (Brown, Chapter 3), the structures and biomechanical functions of the auditory periphery (Echteler, Fay, and Popper, Chapter 5; Rosowski, Chapters 6), and the ways that source features are detected, discriminated, and perceived among mammals (Long, Chapter 2; Stebbins and Moody, Chapter 4). At each level of analysis, mammals, and perhaps all vertebrates, appear to share essentially similar auditory mechanisms and what could be termed a generally "vertebrate" sense of hearing. The comparative auditory research presented here has helped establish the fundamental principles of this vertebrate hearing sense and the biological context within which human hearing will be better understood.

2. The Sense of Hearing Among Mammals

The hearing capabilities of nonhuman mammals are reviewed and discussed in detail by Long (Chapter 2), focusing on the psychoacoustics of nonprimate mammals, by Stebbins and Moody (Chapter 4) on auditory perception in nonhuman primates, and by Brown (Chapter 3) on sound source localization.

2.1 Hearing Sensitivity and Bandwidth

Valid animal psychoacoustic data have been obtained since the 1930s when rigorous animal conditioning techniques were first combined with psychophysical paradigms to determine the audiogram for the chimpanzee (Elder 1934). Since that time, audiograms have been determined and published for over 60 mammal species. Figure 1.1 shows most of these audiograms from a recent review (Fay 1988). In general, conditioning and psychophysical methods have been successfully applied to a wide range of mammals including marsupials, shrews, lagomorphs (rabbits), many rodents, carnivores, pinnipeds (sea lions and seals), insectivores, several families of bats, many primates, ungulates, cetaceans (whales), and a prohoscid (elephant).

With the exception of some burrowing mammals specially adapted for life in tunnels (see Heffner and Heffner 1992), there are three important generalizations that can be made. First, most mammals, including humans, have the

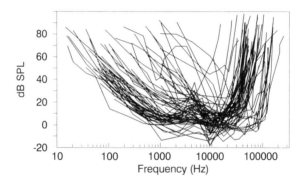

FIGURE 1.1. Behavioral audiograms for over 60 mammals. (Modified after Fay 1988, with permission.)

lowest thresholds within 10 dB of 0 dB SPL (sound pressure level). Rosowski (Chapter 6) has pointed out that the ratio of tympanic membrane to stapes footplate area is constant at about 20 for all mammals investigated. The similarity of best behavioral sound detection thresholds among mammals seems to reflect the similarity in middle ear area ratios and may reflect a fitness advantage for sensitivity to signals as low as, but no less than, irreducible internal and external ambient noise levels (Long, Chapter 2).

Second, the greatest variation among mammals is in the frequency range of hearing. The lower frequency limit of hearing ranges from 20 Hz to 10 kHz (9 octaves) and the upper frequency limit of hearing ranges from 11 kHz to 150 kHz (4 octaves). Some species have essentially nonoverlapping audibility ranges, and no species hears over the entire range to which mammals have adapted.

Third, the highest audible frequency for a given species is negatively correlated with body, head, and ossicle sizes (Rosowski, Chapter 6) and with the distance between the ears (e.g., Heffner and Heffner 1992). It is also reflected in the filtering characteristics of the outer and middle ears as well as the dimensions of the cochlea (Echteler, Fay and Popper, Chapter 5). In general, wide differences in the upper and lower frequency ranges of hearing are consistent with similar diversity in outer, middle, and inner ear dimensions.

The frequency at which a species hears with greatest sensitivity and the frequency range of hearing are due to complementary and coevolved structural adaptations of the middle and inner ears. Rosowski (Chapter 6) has pointed out that sensitivity at high frequencies is associated with outer and middle ear structures that are small and stiff, while sensitivity at low frequencies are associated with large tympanic membrane areas and large, compliant middle ear air spaces. Some species can be classified as specialized for low- or high-frequency hearing on this basis while others (e.g., the cat) exhibit elements of both.

Additional components of this variation are caused by adaptations of the cochlea for the use of particularly high-frequency sound in some echolocating bats and adaptations of the tympanic membrane and middle ear air cavity for detecting particularly low-frequency sound in some rodents. The relatively poor high-frequency hearing of humans is at least in part a consequence of our relatively massive middle ear ossicles and the stiffness of the ossicles and cochlea. It seems unlikely that the human hearing range is a special adaptation for processing speech sounds. Rather, speech production mechanisms have probably adapted to our hearing range.

2.2 Sound Level Processing

The ability to detect a change in sound level is an important hearing function for all animals. Not only does level processing play a role in the perception of source distance (Brown, Chapter 3), but it is also important in binaural and monaural sound localization in azimuth and elevation and the identification of sources through their spectral shapes. Level discrimination also plays a role in detecting signals in noise (Long, Chapters 3; Stebbins and Moody, Chapters 4).

In general, mammals differ somewhat in the lowest level discrimination threshold (LDT). Humans have the lowest thresholds (0.5 to 2 dB), some rodents have the highest thresholds (2 to 6 dB), and other mammals are intermediate. However, in all mammals tested, the LDT tends to be independent of frequency and tends to decline slightly with increasing level. This decline has been termed the "near miss" to Weber's Law (Viemeister 1972). Weber's Law holds that the magnitude of the LDT is proportional to the overall stimulus magnitude. The approximate truth and the generally small deviation from Weber's Law is a conspicuous convergence of function among mammals.

The LDT may be independent of cochlear structure, reflecting more the ways that all nerve fibers encode stimulus level and the ways that nervous systems integrate neural activity (Plack and Viemeister 1993). On the other hand, the hypothesis that sound level processing is based on spatial patterns of excitation on the basilar membrane has not yet been evaluated using comparative data (e.g., Florentine and Buus 1981). At present, there are no indications of specially adapted mechanisms for level discrimination and no reasons to believe that this hearing capacity or its mechanisms differ among mammals in fundamental ways.

2.3 Frequency Processing and Auditory Filters

All mammals analyze simple and complex sounds into their constituent frequency components with varying degrees of resolution. Frequency analysis is used to optimize signal-to-noise ratios for sound detection against a background of competing sounds, to aid in sound source localization in azimuth and elevation, and as a way to identify sound sources according to their spectral shape. Frequency discrimination thresholds indicate the resolution with which simple sound spectra are represented in the nervous system. Masking studies estimating critical masking ratios, critical bandwidths, auditory filter shapes, and psychophysical tuning curves indicate the resolution with which more complex sound spectra are resolved (Long, Chapter 2; Stebbins and Moody, Chapter 4).

The best understood and most important relationship between cochlear structure, mechanics, and hearing performance among mammals is that relating the cochlear frequency-position map for a given species (Echteler, Fay, and Popper, Chapter 5); (von Békésy 1960; Greenwood 1961; 1990) and that species' behavioral performance in frequency analysis (Fay 1992; Long, Chapter 2; Stebbins and Moody, Chapter 4).

Greenwood (1961) measured human critical bandwidths (CB) in psychophysical experiments and then estimated a position-frequency map for the human cochlea using the assumption that CBs (in Hz) represent equal distances on the basilar membrane, independent of center frequency (See Long, Chapter 2, for a detailed discussion of critical bands). Fay (1992) has shown that for humans and most other mammalian species investigated, Greenwood's map function can be fit to empirically determined position-frequency maps using only the length of the basilar membrane and the upper frequency limit of hearing as parameters (See map equations in Echteler, Fay, and Popper, Chapter 5). Furthermore, Fay (1992) demonstrated that behavioral measures of frequency analysis can be reasonably well predicted in several mammalian species using the assumption that critical bands (and other behaviorally-defined measures of frequency resolution) correspond to equal distances along the basilar membrane (about 1 mm per CB for most mammals).

Echteler, Fay, and Popper (Chapter 5) show that mammals may be classified as having generalized or specialized cochleas depending on the fits between empirically-determined and predicted cochlear position-frequency maps. The maps estimated for the mole rat (*Spalax ehrenbergi*) and horseshoe bat (*Rhinolophus ferrumequinum*) appear to be unlike those for other mammals investigated in that they show abrupt slope transitions uncharacteristic of the usual frequency-place maps (Fay 1992; Greenwood 1961; 1990). Those species for which Greenwood's (1990) map function fits empirical position-frequency maps well are termed "cochlear generalists" while those species whose cochlear maps are not well modeled by the Greenwood function (e.g., the horseshoe bat and the mole rat) are termed "cochlear specialists." The relationships between the cochlear map and behavioral frequency acuity demonstrated for humans and other mammals (Fay 1992) suggest that cochlear specializations tend to increase the acuity of frequency analysis at the high frequencies for the horseshoe bat, and at low frequencies for the mole rat.

In general, these considerations suggest that mammal cochleas not only conform to a single structural *bauplan*, but that the mechanical and perceptual functions determined by them are similar among mammals as well. These tentative conclusions help place species differences in cochlear structure, hearing bandwidth, and capacities for frequency analysis in the sort of quantitative biological context that permits not only the development of more useful animal models for human hearing, but also a deeper understanding of the evolution of the ear and the sense of hearing.

2.4 Temporal Processing

As Long (Chapter 2) points out, most natural sounds fluctuate over time in amplitude and in the relative amplitudes of the spectral components present. Yost (1991, 1993) and others have argued that the temporal separations of sounds, their relative temporal onset and offset times, and their relative temporal modulations in amplitude play important roles in determining the perceptual formation of sound source images for human listeners. It seems likely that these are important factors in sound source determination by other species as well. In general, the time patterns of envelope fluctuation are probably important for the detection, identification, and classification of sound sources and play important roles in the perception of species-specific vocalizations (Stebbins and Moody, Chapter 4). In addition, the temporal analysis of a wave-form's fine structure probably plays a role in spectral analysis and in sound localization (Brown, Chapter 3) at frequencies below several kilohertz.

The threshold for detecting a brief sound generally declines as sound duration increases up to several hundred milliseconds in all mammal species studied, as well as in all vertebrates investigated (Fay 1988). This effect can be understood in two ways: (1) stimulus energy (the product of duration and intensity) may be constant at threshold, and (2) sounds of longer duration permit a greater number of independent decisions about their presence (Viemeister and Wakefield 1991). In any case, it appears that temporal integration is determined, at least in part, by the nearly linear way the brain seems to accumulate neural activity in decision making. Thus, there are no reasons to believe that temporal summation is a recently adapted hearing function or one determined by observable differences in cochlear structure. As is true for level processing, temporal integration appears to be determined by the functions of receptors, nerve cells, and simple neural processing that are probably similar among mammals.

When animals are motivated to detect brief, broadband sounds, the time constant of the integrator used in these tasks seems to be at least two orders of magnitude shorter than that revealed in experiments on temporal integration. These minimum integration times have been estimated by measuring the minimum detectable silent gap in an ongoing sound and the temporal modulation transfer function (TMTF). The latter experiment reveals that as modulation frequency increases above a critical value, the minimum detectable modulation depth grows larger. The chinchilla and humans show similar gap thresholds and TMTF shapes. Although comparative data are limited, discrimination between sounds of different durations and repetition rates also suggest only small or insignificant differences in performance between humans and other mammals. These data suggest that these abilities are similar among mammals and are probably determined by the same neural mechanisms, independent of cochlear structure.

2.5 Sound Source Localization

As Brown (Chapter 3) points out, sound source position in azimuth, elevation, and distance is one of the primary perceptual attributes of a sound source. Pumphrey (1950) has argued, in fact, that *the* primitive function of hearing is the location of sound sources.

All vertebrates have paired ears separated by an obstacle (the head) and by a distance that determines interaural time differences. In general, azimuthal localization abilities appear to involve the estimation of interaural correlations in time, overall level, and spectral shape. The relative importance of these variables appears to depend on source frequency and species characteristics such as the frequency range of hearing, head size, and lifestyle. Not all mammals are able to localize sound sources with the same acuity. Although some portion of these differences may be due to head size and the magnitude of the physical cues available, a more important factor seems to be variation in selective pressure for high acuity. Heffner and Heffner (1992) have pointed out that animals having a narrow visual fovea, such as humans and the cat (*Felis catus*), also have small minimum audible angle thresholds, while those animals with a wide foveal "area centralis," such as the domestic cow (*Bos taurus*), have poor directional acuity. This generalization is consistent with the notion that one function of sound localization is to coordinate vision and hearing in the determination of objects in the environment, or at least to help direct the visual gaze toward sound sources.

Many mammals are able to use both interaural time and level cues in azimuthal localization (e.g., primates, cat, and rat). However, some species make very little use of time cues (e.g., hedgehog *Paraechinus hypomelas*), while others make little use of interaural level cues (e.g., horse *Equus caballus*). The burrowing pocket gopher (*Geomys bursarius*) seems unable to make use of any binaural cues (Heffner and Heffner 1992).

Humans use monaural, or pinna, cues for azimuthal sound localization (e.g., Butler 1986). Although pinnae are directional receivers in many mammals (Rosowski, Chapter 6), there are little data on monaural localization abilities in nonhuman mammals. Heffner and Heffner (1984) have demonstrated that the horse fails at front-back discriminations (where monaural cues are probably necessary) when the frequency range of the sound source was restricted to 500 Hz and below. This suggests that horses can make use of monaural cues only for broadband stimuli extending to high frequencies (8 kHz and above).

Most mammals tested can localize broadband sources in elevation with an acuity of 2 to 13 degrees (Brown, Chapter 3). It is likely that the useful cue here is the elevation-dependent spectral filtering characteristics of the pinnae (Rosowski, Chapter 6). This illustrates dramatically that capacities for frequency analysis are required not only for source identification, but also for monaural localization and for directional hearing in the vertical plane.

True, three-dimensional sound localization requires an estimation of source distance. Few studies have investigated abilities for distance estimation in humans or other mammals. Brown (Chapter 3) provides a list of potential monaural and binaural cues that could help determine sound source distance. These include knowledge of the source levels of familiar sounds (image constancy), distance-dependent high-pass filtering by the environment, and distance-dependent degrees of reverberation and wave-form distortion. Studies of auditory image formation in three dimensions will likely increase in the future.

In order to develop the best animal models for human sound source localization, it must be recognized that there are relatively wide differences among mammals in localization acuity, in the effective cues used, and in the central systems that are most important for processing these cues. These differences may not be predictable from ear structures or from peripheral neurophysiological data and may require behavioral experiments for their definition.

2.6 Complex Sound Perception

Most animal psychoacoustic research has used relatively simple sounds such as tones and noise in order to quantitatively describe the fundamental hearing capabilities in a systematic way. Except for the bandwidth of hearing and aspects of frequency analysis that are determined in large part by outer, middle, and inner ear structures, most fundamental hearing capabilities tend to be similar among mammals, reflecting the conservative nature of peripheral sound encoding. Most natural sound sources produce complex, broadband sounds having rapid fluctuations in amplitude, frequency, and bandwidth. In an attempt to study the perception of these complex sounds, the researcher is faced with choosing stimuli to study from what appears to be an infinite number of possible, arbitrary sounds. Is there a systematic and rational way to begin to study the perception of complex sounds?

One approach to this question focuses on sets of complex sounds that are known to have specific relevance or "biological significance" to the animal: vocalization and communication sounds. In Chapter 4, Stebbins and Moody review several experiments on the perception of inter- and intraspecific communication sounds in monkeys. Based on field studies that have identified the conditions of occurrence and the possible meaning of certain vocalizations, specific questions are raised. These include the question of whether monkey communication and human speech perception are based on the same principles or processes and the question of whether monkeys perceive human speech as we do. These questions seem to be the obvious ones in this context, and they redirect the focus from questions of what these animals can discriminate if forced to do so to questions of the basis for their categorization and classification of sounds. The theoretical underpinning for such studies comes from research on human speech perception.

After reviewing studies on classification, categorization, laterality of function, and human speech perception by monkeys, Stebbins and Moody (Chapter 4) conclude that monkeys as well as humans show evidence of perceptual constancy, categorical perception, hemispheric laterality, and the salience of frequency modulation as an important feature in communication signals. This provides quantitative evidence for cognition and cognitive strategies in these animals, properties often reserved only for humans. As Stebbins and Moody point out, these studies offer a relatively untapped, quantitative view into the possibility of animal awareness and consciousness. The conclusion is reached that human speech has its origins in the communication signal processing in other species.

These studies illustrate a general problem for comparative hearing research. There seems to be a rather wide gap between the tones and noises often used in animal psychoacoustics and the "biologically relevant," species-specific communication and orientation sounds often used in experiments inspired by ethological studies. The focus in ethology is often on species differences and on what is "special" or unique to a species. By definition, species-specific communication sounds are unique. Does this mean that studying "biologically relevant" complex sound perception in animals requires that we treat each species as a unique problem using unique stimuli? If not, is there really anything of fundamental importance to be learned by studying a given species using the unique communication sounds of another species? How can we get at general principles here?

There may be a middle ground—a way to investigate the perception of complex sounds, not as a euphemism for "communication sounds," but as the characteristics of all the sound sources common in the natural world. As suggested at the beginning of this chapter, the most general functions of hearing are to identify, classify, and locate sound sources within an auditory scene. In most hearing experiments, the sources are loudspeakers or earphones broadcasting signals that have been specified, using the vocabulary of linear acoustics: frequency, phase, amplitude, bandwidth, spectrum, etc. These are not characteristics of sources but of sounds. Imagine a more ecological sort of acoustics in which the variables are characteristics of materials —size, shape, stiffness, mass, driving forces, modes of vibration, etc.—the things that make sources such as the vocal tract or a footfall "sound" as they do. Perhaps these variables could be a basis for a new psychoacoustics of sound sources that focused on the abilities of humans and other animals to perceive a scene made up of one or more of them (e.g., Freed 1990). One could ask whether footfalls are classified similarly in spite of wide variations in linear acoustic characteristics, or whether the ability to locate flowing water is degraded by omnidirectional wind noise, or whether two ears are required to detect, identify, and locate one source among many in a natural scene.

Although this proposal for future research is rather vague at this point, an experimental and conceptual focus on sources rather than sounds may help advance our understanding of the functions of hearing.

3. Opportunities for Future Research

Future studies in comparative auditory research could perhaps address some of the following problems and questions raised in this chapter:

1. On what qualitative and quantitative dimensions do humans differ from other mammals in the sense of hearing? Small differences or those that are determined by ear structure still need to be investigated quantitatively. However, larger questions concern whether and to what extent nonhumans perceive or determine sound sources as humans do.
2. New research programs need to be developed that study the sense of hearing and not just the response to sound. These paradigms should be applicable to many species and not uniquely tied to species-specific characteristics.
3. Stronger conceptual and experimental links between behavioral and neurophysiological studies need to be made. Similarly, we also need stronger conceptual and experimental links between human and animal psychophysics and between the sense of hearing and structures of the ear and brain.
4. What are the functional meanings of diverse ear and brain morphologies? Differences in structure do not necessarily lead to different functions.
5. The field would be advanced by a comprehensive, detailed analysis of several carefully selected species with the aim of understanding general principles. The general principles will provide a context for understanding species differences and specializations or adaptations. At the same time, wide-ranging exploration of many species should continue so that we can more fully understand the biological context within which human hearing evolved.

Acknowledgments. The writing of this chapter was supported by a Center Grant from the National Institutes of Health, National Institute of Deafness and other Communicative Disorders to the Parmly Hearing Institute.

References

Békésy G von (1960) Experiments in Hearing. New York: McGraw-Hill.
Bregman AS (1990) Auditory Scene Analysis: The Perceptual Organization of Sound. Cambridge, MA: MIT Press.

Butler RA (1986) The bandwidth effect on monaural and binaural localization. Hear Res 21:67–73.

Elder JH (1934) Auditory acuity of the chimpanzee. J Comp Physiol Psychol 17:157–183.

Fay RR (1988) Hearing in Vertebrates: A Psychophysics Databook. Winnetka, IL: Hill-Fay Associates.

Fay RR (1992) Structure and function in sound discrimination among vertebrates. In: Webster DB, Fay RR, Popper AN (eds) The Evolutionary Biology of Hearing. New York: Springer-Verlag, pp. 229–263.

Florentine M, Buus S (1981) An excitation pattern model for intensity discrimination. J Acoust Soc Am 70:1646–1654.

Freed DJ (1990) Auditory correlates of perceived mallet hardness for a set of recorded percussive sound events. J Acoust Soc Am 87:311–322.

Glasberg BR, Moore BCJ (1990) Derivation of auditory filter shapes from notched-noise data. Hear Res 47:103–138.

Green DM, Swets JA (1966) Signal Detection Theory and Psychophysics. New York: John Wiley and Sons.

Greenwood DD (1961) Critical bandwidth and the frequency coordinates of the basilar membrane. J Acoust Soc Am 33:1344–1356.

Greenwood DD (1990) A cochlear frequency-position function for several species— 29 years later. J Acoust Soc Am 87:2592–2605.

Heffner HE, Heffner RS (1984) Sound localization in large mammals: Localization of complex sounds by horses. Behav Neurosci 98:541–555.

Heffner RS, Heffner HE (1992) Evolution of sound localization in mammals. In: Webster DB, Fay RR, Popper AN (eds) The Evolutionary Biology of Hearing. New York: Springer-Verlag, pp. 691–716.

Kraus N, McGee T (1992) Electrophysiology of the human auditory system. In: Popper AN, Fay RR (eds) The Auditory Pathway: Neurophysiology. New York: Springer-Verlag, pp. 335–403.

Liberman MC (1982) The cochlear frequency map for the cat: Labeling auditory nerve fibers of known characteristic frequency. J Acoust Soc Am 72:1441–1449.

Moore BCJ (1993) Frequency analysis and pitch perception. In: Yost WA, Popper AN, Fay RR (eds) Psychoacoustics. New York: Springer-Verlag, pp. 56–115.

Myrberg AA (1981) Sound communication and interception in fishes. In Tavolga WN, Popper AN, Fay RR (eds) Hearing and Sound Communication in Fishes. New York: Springer-Verlag, pp. 395–426.

Plack CJ, Viemeister NF (1993) Suppression and the dynamic range of hearing. J Acoust Soc Am 93:976–982.

Popper AN, Fay RR (1992) The Auditory Pathway: Neurophysiology. New York: Springer-Verlag.

Pumphrey RJ (1950) Hearing. Symp Soc Exp Biol 4:1–18.

Rogers PH, Lewis TN, Willis MJ, Abrahamson S (1989) Scattered ambient noise as an auditory stimulus for fish. J Acoust Soc Am 85:S35 (abstract).

Ruggero M (1992) Physiology and coding of sound in the auditory nerve. In: Popper AN, Fay RR (eds) The Mammalian Auditory Pathway: Neurophysiology. New York: Springer-Verlag, pp. 34–93.

Serafin SV, Moody DB, Stebbins WC (1982) Frequency selectivity of the monkey's auditory system: Psychophysical tuning curves. J Acoust Soc Am 71:1513–1518.

Viemeister NF (1972) Intensity discrimination of pulsed sinusoids: The effects of filtered noise. J Acoust Soc Am 51:1265–1269.

Viemeister NF, Wakefield GH (1991) Temporal integration and multiple looks. J Acoust Soc Am 90:858–865.

Webster DB (1992) Epilog to the conference on the evolutionary biology of hearing. In: Webster DB, Fay RR, Popper AN (eds) The Evolutionary Biology of Hearing. New York: Springer-Verlag, pp. 787–794.

Webster DB, Fay RR, Popper AN (1992) The Evolutionary Biology of Hearing. New York: Springer-Verlag.

Webster DB, Popper AN, Fay RR (1992) The Mammalian Auditory Pathway: Neuroanatomy. New York: Springer-Verlag.

Yost WA (1991) Auditory image perception and analysis: The basis for hearing. Hear Res 56:8–18.

Yost WA (1993) Overview of psychoacoustics. In: Yost WA, Popper AN, Fay RR (ed) Psychoacoustics. New York: Springer-Verlag, pp. 1–12.

Yost WA, Popper AN, Fay RR (1993) Human Psychophysics. New York: Springer-Verlag.

2
Psychoacoustics

GLENIS R. LONG

1. Introduction

While there are now large bodies of research in both human psychoacoustics (reviewed in Vol 3 of this series) and nonhuman mammalian auditory physiology and anatomy (reviewed in Vols 1 and 2), the understanding of mammalian hearing has been handicapped by a failure to integrate these two areas of research. Many human psychoacoustic papers attempt to account for, and model, human psychophysical performance based on some convenient selection of nonhuman mammalian anatomical and physiological measures of hearing, without determining whether the animals perceive sounds in the same way as the human subjects. Although the mammalian auditory system has several components in common across species, there can be extreme differences when the auditory system is modified due to an animal's specialized use of sound (see Echteler, Chapter 5). For example, the cochlea of the greater horseshoe bat (*Rhinolophus ferrumequinum*) departs significantly from that of other species (reviewed in Pollak and Casseday 1989), and estimates of frequency resolution (Long 1977, 1980a,b) and frequency discrimination (Heilmann-Rudolf 1984) from this species reflect these differences. These specializations are related to the use of sound for echolocation. Humans also use sounds in a very sophisticated way in speech perception, and it is probable that the human auditory system is also specialized (especially in the frequency region associated with speech). Other species use sounds to communicate (see Stebbins and Moody, Chapter 4), but the nature of the sounds used by other species are not as rich as human speech and the content conveyed is not as sophisticated.

The only neurophysiological measures obtained from humans are remote evoked-response recordings that are obtained by placing electrodes on the surface of the head (reviewed by Kraus and McGee 1992). Evoked responses are limited in specificity and cannot be used to test hypotheses about the relationship between perception and neurophysiology. Psychoacoustic research on the species used in anatomical and physiological research is, therefore, the only currently available tool to provide a clear understanding of the dependence of perception and sensation on anatomy and physiology. Furthermore, research on the same species can only reach its full potential if similar stimuli are used in both the psychoacoustic and physiological research.

Comparative mammalian psychoacoustics not only helps us understand human auditory perception but also provides essential information about the potential importance of acoustic stimuli to each species. Different species in the same physical environment will be able to respond to a limited subset of the acoustically available sounds. Even species that respond to the same range of sounds may be able to discriminate different aspects of the sounds. An understanding of the role of acoustic cues in predator/prey relationships and communication depends on an understanding of the ability of each species to process the acoustic events in the environment. If a better understanding of the relationship between sensation and the anatomy and physiology of the auditory system is obtained, it will be much easier to predict auditory processing skills in mammals from their anatomical and physiological structures. Until there is that understanding, any discussion of the acoustic environment of each species must be based on behavioral research using that species.

Since most psychoacoustic research has been conducted using human subjects, much of this chapter will compare research with humans with research with nonhuman mammals. In order to facilitate communication, the word "animal" will be used to refer to nonhuman mammals. It is not possible to summarize all the psychoacoustic research completed on animals. The research up to 1988 is available in the invaluable databook compiled by Fay (1988). This chapter will provide an overview of the comparative mammalian psychoacoustics of simple sounds. Research on binaural hearing is covered by Brown (Chapter 3) and many aspects of complex sound perception are covered Stebbins and Moody (Chapter 4).

2. Methodology

It is not easy to obtain reliable and valid behavioral measures of hearing from animals. Adult human psychoacoustic research is based on much experience with psychophysical paradigms. Current paradigms are based on research on decision making and evaluate sensory processing so that the effects of past experience, emotional meaning of the sounds, and response bias are minimized (for a good review of current procedures see MacMillan and Creelman 1991). These experimental paradigms are computer controlled and all that is needed is a combination of verbal instructions to the subject and practice with the experimental task. In those cases in which an alternative cue becomes available to the subject, the experimenter may be informed by the subject that something is wrong (or the experimenter will detect the alternative cue when they listen to the experimental task).

2.1 Training

Psychoacoustic research with animals cannot use many of the paradigms developed for use with adult humans. The animal must indicate to the experimenter that it detects the required stimulus contrast (reviewed in Stebbins 1990). This can be done by training the animal using classical or operant conditioning or by relying on an unconditioned reflex to the sound. However, the use of reflexes can be criticized. A failure to respond may not mean that the stimulus is not detected by the animal, since a specific stimulus may be detected but may fail to evoke the reflex under investigation. But, reflex modification by a prior stimulus appears to be a useful technique (reviewed in Hoffman and Ison 1992). Although the implementation of computers in animal research has helped make the research more objective, some animal research cannot easily be placed completely under computer control because of the nature of the optimal response or the reinforcement for that species (e.g., the use of live mealworms as a reward cannot be under computer control because they refuse to stay still).

The type of response demanded from a given species can have a dramatic impact on the time needed to train an animal. Traditional learning theory (see Mackintosh 1974) would suggest that all training procedures would work equally well with all species. If this is true, the best approach to training a new species would be to imitate research with other species (usually the methods used with a rat or a chinchilla). Recent research on learning suggests that there is a need to look at an animal's normal responses to the stimuli and the reinforcer or reward (reviewed in Davey 1989). For example, if the animal is normally a forager for food, it makes sense to give it a task (response) that involves active food seeking. One can speed up training significantly by selecting the appropriate response (see Harrison 1992). The dependence of animal psychoacoustics on classical learning theory has led investigators to motivate an animal by food or water deprivation (e.g. "starve the animal to 80% of normal weight") before starting the research. This may not be optimal for all species or conditions (reviewed in Moran 1975). There are some conditions in which deprivation does not serve as a motivator and may cause the animal to get sick.

Even when the animal is responding consistently, the experimenter must ensure that it is responding to the stimulus contrast of interest to the experimenter. The animal may learn to respond to cues that are not those anticipated by the experimenter. One example of this may be the low-frequency thresholds obtained from the bats *Myotis oxygnathus* and *Rhinolophus ferrumequinum* by Ayrapet'yants and Konstantinov (1974). While the thresholds above 10 kHz are consistent with audiograms from similar species (reviewed in Long and Schnitzler 1975), thresholds below 10 kHz are independent of frequency and much lower than those obtained by other researchers. Different groups of bats were used to obtain these two portions of the audiogram. The first group of bats was trained at high frequencies but failed to respond to the low-frequency tones. Another group of bats was trained using low-frequency tones. The low-frequency part of the audiogram comes from these animals. The tonal stimuli were generated without a gradual rise and fall of stimulus level. Consequently, an animal could have learned to respond to the clicks generated at stimulus onset (these clicks should be relatively frequency independent). Animals trained at high frequencies would have been reinforced for responding to the tones and not for responding to the clicks. They would, therefore, not respond to the low frequencies where they would only hear the clicks.

Other artifacts such as adjustment of an attenuator at the beginning of the trial, also can be used by the animals. These cues may not be audible to the experimenter. An animal may learn to ignore such cues when the experimental contrast is very salient, only to start using them once the task becomes more difficult. The possibility that the animal may be responding to cues at frequencies other than those planned by the experimenter (such as onset transients or harmonic distortion) is increased if there is cochlear damage restricting the range of frequencies that an animal can hear (e.g. Prosen, Halpern, and Dallos 1989) and the stimuli presented are no longer easily detected by the animal.

2.2 Attention

If the animal is not attending to the stimulus on all trials, the assumptions underlying the threshold estimate may not be valid. One can minimize inattention by using experimental paradigms requiring the animal to make an observing response, such as pressing a key when it is ready to work. Psychoacoustic research with human infants has many of the same problems. One cannot give the infant verbal instructions or be sure that the infant is attending to the stimulus each time that it is presented. It is also possible that the infant is not equally motivated to make a response under all experimental conditions. The effects of these variables in infant psychoacoustics have been reviewed in Bargones, Marean, and Werner (1992), Viemeister and Schlauch (1992), and Wightman and Allen (1992). It is possible to obtain very reliable psychoacoustic data from both infants and nonhuman mammals as long as one is prepared to invest the time and care needed. The consistency between many of the measures from different labs testify to this (see Fay 1988).

2.3 Threshold Determination

Threshold estimations in animal research are frequently based on outdated concepts of threshold. Most contemporary human research is based on signal-detection theory (reviewed in MacMillan and Creelman 1991). Signal-detection theory assumes that an individual is comparing overlapping distributions of sensory events. Each time a stimulus is presented, the sensation varies so that the sensation evoked by many presentations of a single stimulus will have a Gaussian distribution. Similar stimuli will thus have overlapping distributions. False alarms (responding in the absence of stimuli) can only be reduced if the number of responses to the stimulus (hits) is also reduced. An optimum response strategy would thus produce false alarms in a significant proportion of the trials. In the absence of false alarms, it is not possible to determine sensitivity uncontaminated by response bias. Unfortunately most of the psychoacoustic research using animals assumes that any false alarm is an indication of a lack of stimulus control. Often, false alarms are punished, producing conservative response strategies which will elevate thresholds. Using an adjusted percent correct, in which false alarms are assumed to be an indication of error, is also not valid (see MacMillan and Creelman 1991). Ideally, one would use a procedure such as d' that allows the use of false alarms to estimate sensitivity (see MacMillan and Creelman 1991), independent of the animal's willingness to respond (criterion or response bias). An alternative would be the use of a paradigm that forces the animal to choose between two alternatives (two alternative forced choice or 2AFC; see MacMillan and Creelman 1991). Unfortunately d' measures are too time-consuming for animal research and it is difficult to train an animal to make 2AFC responses (reviewed in Burdick 1979; Harrison 1992). Constraints of time and training procedures often make these impossible. One can avoid many pitfalls by monitoring false alarms and keeping them at a constant level (between 10% and 20%).

2.4 Stimulus Presentation and Calibration

The stimulus presentation paradigm can also have a dramatic effect on an animal's the ability to learn a discrimination task (reviewed in Burdick 1979; Harrison 1992). Human discrimination performance is also modified by the stimulus presentation paradigm (e.g., Jesteadt and Sims 1975; Viemeister and Bacon 1988; Turner, Zwislocki, and Filion 1989). The tasks most commonly used with human subjects are those tasks that both give the largest thresholds (least sensitive) and are the most difficult for the nonhuman mammals. Consequently, unless care is taken to compare results with similar stimulus paradigms, the least sensitive measures of human performance are being compared with the most sensitive measures from nonhuman species.

In most human psychoacoustics, stimuli are presented using headphones. This enables exact specification of the acoustic characteristics of the stimuli at the eardrum (at least for frequencies up to 4 kHz) by calibrating the headphones using a coupler that has the same volume of air as that normally enclosed between the headphones and the average human eardrum (see Stinson and Lawton 1989). Most researchers use headphones with a very flat frequency response up to 4 or 5 kHz but that rapidly become less efficient at higher frequencies. Consequently, the broadband stimuli used in many human experiments are limited in frequency range, a factor that may influence the results (see Section 6.2).

In contrast, most nonprimate research is done in a free field, i.e., the sound is presented through loudspeakers. In free-field experiments, the stimuli are calibrated by placing a microphone at the place normally occupied by the subject's head. There are consistent differences between free-field and earphone measurements in humans. These differences are partly due to the resonance characteristics of the ear canal (coupler calibration estimates the level at the eardrum and free-field calibration estimates the sound pressure before the resonance of the outer ear). The use of two ears rather than one also influences the results (Killion 1978).

In animal research, the sound is delivered through speakers some distance from the animal. Reflections from the environment and parts of the cage may modify the signal (especially at high frequencies), leading to large differences in the stimulus level at different parts of the cage. If the animal's head is fixed in position during stimulus presentation (i.e., in those conditions requiring the animal to make an observing response), calibration is relatively easy. The calibrating microphone is placed in the position the animal's head usually occupies during the observing response, taking care to correct for the effect of the animal in the sound field. In other paradigms when the animal is free to move in the cage during stimulus presentation (e.g., conditioned avoidance), the stimulus is specified by using the mean value obtained when a microphone is placed in several positions in the cage. Animals often choose relatively constant positions in the test environment during testing, which may reflect an attempt to optimize stimulus detection. In such cases, the mean calibration value would tend to underestimate thresholds (Heffner and Heffner 1991). The increased probability that the animal may not be attending to the sound when it is not required to make an observing response would also tend to underestimate thresholds in these paradigms (see Section 2.2).

Background noise can be a problem if the animal has to be tested in a more natural environment. Low-frequency sensitivity in the elephant (*Elephas maximus*; see Fig. 2.1) may have been underestimated between 31.5 and 2000 Hz since thresholds in this region are similar to the level of the background noise in the testing environment at these frequencies (Heffner and Heffner 1982). Careful measurement of the background noise enables interpretation of the results.

2.5 Scaling

Evaluating qualitative aspects of sounds (scaling) is difficult. Reaction time has been used as an estimate of loudness (e.g., Moody 1970). It is assumed that the shorter the reaction time, the louder the stimulus. Confusion between different stimuli, as reflected in reaction time measurement, can be used as an indication of the similarity between the two stimuli. Similarity scaling using reaction time has been developed for birds by Dooling (reviewed in Dooling 1989; Stebbins 1990).

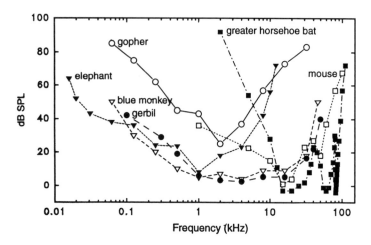

FIGURE 2.1. Behavioral audiograms from mammals of varying size. The threshold (in dB Sound Pressure Level, SPL) at each frequency varies with the animal's size unless there is specialization of the outer & middle ear. Blue monkey *Cercopithecus mitus* (open triangles, Brown and Waser 1984). Indian elephant *Elephas maximus* (filled triangles, Heffner and Heffner 1982). Pocket gopher *Geomys bursarius* (open circles, Heffner and Heffner 1990). Gerbil *Meriones unguiculatus* (filled circles, Ryan 1976). Mouse *Mus musculus* (open squares; Ehret 1974). Greater horseshoe bat *Rhinolophus ferrumequinum* (filled squares, Long and Schnitzler 1975).

2.6 Summary

Much care must be taken in the choice of experimental paradigms to ensure that differences in psychoacoustic performance between species reflect differences in sensory processing. Ideally, one would use the same procedure and environment for all species. When this is not possible, the performance in each environment should be compared to a known quantity such as the performance of humans in the same environment. In some instances, the same human subjects (the researchers) are used in many experiments (e.g., Sinnott, Brown, and Brown 1992). Furthermore, comparison of different psychoacoustic observations within a species must be made with care unless the data were obtained with similar stimulus presentation and threshold estimation procedures.

3. Hearing Sensitivity

With the possible exception of the burrowing animals, such as the pocket gopher (*Geomys bursiarius*) and mole rat (*Spalax ehrenbergi*; reviewed in Heffner and Heffner 1990), the lowest level of stimuli that can be detected within the best frequency range is approximately the same for all mammalian species. If mammals were much more sensitive to sound, they would be able to detect the random motion of the air molecules (Sivian and White 1933).

There are, however, large differences in the range of frequencies that can be detected by different mammalian species. Some examples of audiograms obtained from different species are given in Figure 2.1. To a large extent, the frequency range is determined by the properties of the outer and middle ears, which are correlated with head size. The impedance mismatch between the sound conduction medium (air) and the fluid in the cochlea needs to be overcome for sound to be detected. The structures of the outer and middle ears amplify the sound pressure to partially compensate for this impedance mismatch. Consequently, the physics of the outer and middle ears (reviewed in Relkin 1988; Rosowski, Chapter 6) is the major determinant of the range of frequencies detected. In the absence of special modifications, the size of the outer ear, middle ear, and ear drum and the mass of the middle ear bones all depend on the size of the head. Consequently, the transfer function of the outer and middle ears also depends on head size (for a review see Rosowski 1992). Most marine mammals have special adaptations to overcome the impedance mismatch (reviewed by Ketten 1992). The mole rat also appears to have special modifications for detecting substrate-borne vibrations rather than airborne vibrations (Rado et al. 1989). Similar modifications for detecting substrate vibrations may explain the poor sensitivity of the burrowing pocket gopher (*Geomys bursiarius*) to airborne sound (see Fig. 2.1).

The acoustics of sound generation and propagation (reviewed in Griffin 1971; Wiley and Richards 1978; Michelsen 1992) may lead some species to use communication sounds or prey detection cues outside the frequency range consistent with their head size. These animals show specializations of the auditory system (outer, middle, and inner ears) to expand the audible range. Some small mammals who live in environments demanding enhanced low-frequency detection have developed modified middle ears that extend their hearing to permit the use of low-frequency warning signals (reviewed in Webster and Plassman 1992). For example, both the desert-dwelling kangaroo rat (*Dipodomys merriani*) and the steppe-dwelling gerbil (*Meriones unguiculatus*) need to transmit warning signals that are not easily located and must carry long distances. The extreme specialization of the outer and middle ears of *Meriones unguiculatus* means that these organs take up approximately two-thirds of the cross-section of its head (Ravicz, Rosowski, and Voigt 1992). In addition, the extreme low frequencies detected by the elephant may be correlated with the use of infrasound for communication (Payne, Langbauer, and Thomas 1986; Poole et al. 1988).

Regions of enhanced sensitivity in the audiograms (see Fig. 2.1) appear to be associated with the use of narrowband sounds for communication (e.g., *Mus musculus*; discussed in Ehret 1989) or the use of echolocation sound with relatively long, constant frequency components (e.g., the greater horseshoe bat *Rhinolophus ferrumequinum*; Long and Schitzler 1975).

It has been suggested that the need to localize sound is the major determinant of the highest frequency detected (reviewed in Heffner and Heffner 1992). This discussion is based on traditional views of auditory localization based on interaural judgments of tonal stimuli. It ignores recent evidence that cross-spectrum cues (both monaural and binaural) stemming from the characteristics of the external ear (particularly the pinna) play a large role in the localization of complex signals (reviewed in Rice et al. 1992).

4. Frequency Processing

Research into an animal's ability to process the frequency characteristics of sound can investigate either the ability to detect that a sound has changed in frequency (frequency discrimination) or the ability to extract one sound from a background of other sounds (frequency resolution).

4.1 Frequency Discrimination

Although cross-species comparisons are limited unless similar paradigms are used in all species, some trends are clear despite differences in methods. As Figure 2.2 shows, it is clear that estimates of frequency discrimination ability in human beings are significantly lower than for rodents (Fay 1974; Long 1983; Long and Clark 1984). Monkeys and cats tend to fall in between (Fay 1974; Sinnott, Brown, and Brown 1992). The only species with frequency discrimination thresholds approaching that of human beings (see Figs. 2.2 and 2.4) are the dolphin *Tursiops truncatus* (e.g., Thompson and Herman 1975) and the greater horseshoe bat *Rhinolophus ferrumequinum* (Heilmann-Rudolf 1984). Aspects of the experimental paradigms that may limit across species comparison are discussed below.

Frequency discrimination appears to be a difficult task to learn (reviewed in Prosen et al. 1990a) and the animals investigated will use any other cues available. This means than when one is testing in a frequency region where the audiogram slopes steeply (or the sound delivery system changes rapidly), the animal may use loudness differences. Pfingst (e.g., Pfingst and Rush 1987; Pfingst and Rai 1990) has developed a paradigm which avoids the problem and independently assesses the effects of level and frequency changes. This method simultaneously provides level and frequency difference limens.

Human studies normally use two-interval forced-choice paradigms to estimate the minimal detectable frequency change (frequency difference limen, dlf). The subject has to indicate which of two sequential periods, each containing a tone burst, contained the higher frequency sound (e.g., Wier, Jesteadt, and Green 1976). Nonhuman mammals do not respond consistently when given two-interval forced-choice tasks (reviewed in Burdick 1979). They do, however, respond well to a change in the frequency of a series of constant frequency background pulses.

An alternative way to evaluate frequency discrimination requires the sub-

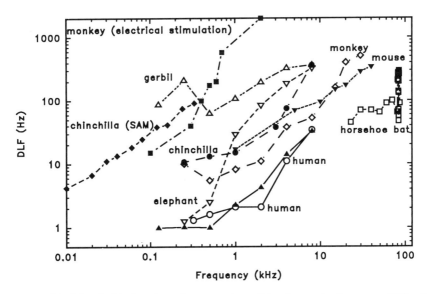

FIGURE 2.2. Thresholds for the minumum detectible change in frequency (dlf) of pulsed tones from a range of species. Chinchilla *Chinchilla langier* (filled circles, Nelson and Kiester 1978). Indian elephant *Elephas maximus* (open downward triangles, Heffner and Heffner 1982). Human *Homo sapiens* (open circles, Fastl 1978; filled triangles, Sinnott, Brown, and Brown 1992); monkey Macaca sp. (diamonds, Stebbins 1973). Gerbil *Meriones unquiculatus* (open triangles, Sinnott, Brown, and Brown 1992). Mouse *Mus musculus* (filled downward triangles, Ehret 1975). Greater horseshoe bat *Rhinolophus ferrumequinum* (open squares, Heilmann-Rudolf 1984). Discrimination thresholds for two tasks based on temporal processing of frequency are included for comparison. Discrimination of changes in the rate of SAM noise from chinchilla *Chinchilla langier* (filled diamonds, Long and Clark 1984). Discrimination of changes in the frequency of electrical stimulation of the cochlea in macaque monkeys, *Macaca* sp (filled squares, cited in Pfingst 1988).

jects to detect a modulation of the frequency of a continuous tone (frequency modulation or FM). This task may be more "natural" for the animal, since many natural sounds vary in frequency (reviewed in Gould 1983; Brown, Chapter 3). When the performance of human subjects on both pulsed-tone discrimination and FM detection are compared (e.g., Jesteadt and Sims 1975; Fastl 1978), very different results are obtained with the two methods. Frequency discrimination threshold patterns obtained with the different paradigms can be seen in Figure 2.3. The pattern of results is very similar for both humans and chinchillas (Nelson and Kiester 1978; Long and Clark 1984). When low-frequency tones are discriminated, the subjects find it easier to detect the differences in the pulsed tones. The difference between the two methods is small at high frequencies. The two paradigms give essentially identical results in the mouse (which can only hear the higher frequencies). In

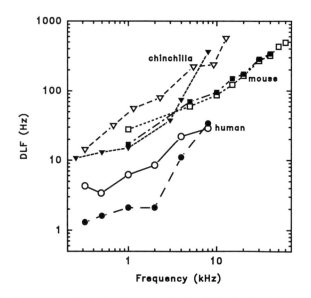

FIGURE 2.3. Frequency discrimination thresholds (dlf) obtained from experiments using pulsed tones (filled symbols) or frequency-modulated tones (open symbols). Chinchilla *Chinchilla langier* (triangles, pulsed tone, Nelson and Kiester 1978; FM, Long and Clark 1984). Mouse *Mus musculus* (squares, Ehret 1975). Human *Homo sapiens* (circles, Fastl 1978).

the middle of the range of frequencies detected by each species, the frequency difference needed for discrimination can be approximated by Weber's Law (the change in frequency is a constant ratio of the comparison frequency).

4.1.1 Potential Neurophysiological Bases for Frequency Resolution

There are two potential mechanisms for coding frequency in the auditory nerve (e.g., Ruggero 1992). The first is rate/place coding in which the frequency of the tone is coded by the rate of firing of different neurons innervating different regions of the cochlea. The second uses the period between the spikes (temporal coding) produced by the phase locking of the spikes to a fixed phase of the stimulus. Javel and Mott (1988) provided evidence that one can more parsimoniously account for many aspects of frequency discrimination performance by assuming that the decision is based on the temporal characteristics (synchronized activity) of the auditory nerve rather than by the rate/place code. However, the minimal frequency difference which can discriminated appears to depend on the length of the basilar membrane (e.g., Prosen et al. 1990a; Fay 1992). This pattern supports the notion that place coding is important in normal hearing animals. Animals with cochlear pathology (who would receive less specific place information) appear to use periodicity (Prosen and Moody 1991). It is possible that normal hearing animals use whichever cue is optimal in each task.

4.1.2 Frequency Discrimination Based on Temporal Cues

Frequency discrimination is possible even when place coding is not possible. Sinusoidally amplitude-modulated white noise (SAM noise) stimulates all areas of the cochlea equally but is perceived by humans as having a pitch determined by the modulation rate (e.g., Burns and Viemeister 1981). The spectrum of this stimulus is constant and the pitch is believed to be based on the temporal aspects of the stimulus. Thus, the detection of the change of modulation rate provides an estimate of the ability of an animal to use temporal cues. Chinchillas (see Fig. 2.2) are able to do this task but need larger changes in rate than human subjects tested using the same paradigm (Long and Clark 1984). However, the chinchilla's performance on this task is more similar to the human subject's performance than is their performance on FM detection.

Electrical stimulation of the cochlea with a single-channel implant stimulates all nerves equally, restricting the animal or human to temporal coding of frequency. Although humans perform better than other primates when discriminating changes in the frequency of acoustic stimuli (reviewed in Pfingst 1988; Sinnott, Brown, and Brown 1992), the differences disappear when the stimuli are generated by single-channel implants (see Fig. 2.2; reviewed in Pfingst 1988). Although SAM noise and electrical stimulation of the cochlea both force the animal to rely on periodicity cues, there are differences in the ability of an individual to discriminate changes in both types of stimuli (Pfingst and Rai 1990). These differences may stem from limitations in the ability of the animal to detect the temporal characteristics of the SAM noise (see Section 6.2), which limits performance in the SAM task.

4.2 Frequency Resolution

Frequency resolution refers to the ability to detect a stimulus at one frequency in the presence of sounds at other frequencies. Frequency resolution determines many aspects of the processing of complex sounds in humans (reviewed in Moore 1986). For example, hearing-impaired subjects who have impaired frequency resolution have difficulty discriminating speech in noise (reviewed in Moore, Glasberg, and Simpson 1992). If improved frequency resolution is correlated with an increased ability to separate a signal from the background noise, one might expect that animals which are more dependent on sound analysis for survival would have improved frequency resolution. Bats and dolphins, who use sounds to echolocate, have very good neural frequency resolution and also give narrower psychoacoustic estimates of frequency resolution than other species. Frequency resolution is especially good in those species of bats (e.g., greater horseshoe bats such as *Rhinolophus fer-rumequinum*) who use echolocation sounds with long, constant frequency components. They compensate for the doppler shift in the echo produced by flight to keep the echo to be analyzed in a narrow frequency range (reviewed in Pollak and Casseday 1989).

A concept central to psychoacoustic research on frequency resolution is the "critical band" (CB). This concept (first outlined by Fletcher 1940) stems from evidence that many, very different psychoacoustic tasks yield results that indicate sounds within a band of frequencies seem to be processed in a manner consistent with an interpretation of acoustic summation within a single filter. Each one of these filters is called a critical band. Stimuli that fall in separate critical bands do not follow acoustic rules. Evidence for a peripheral filter, such as the critical band, comes from research into masking loudness of complex sounds, sensitivity to relative phase, tonal dissonance, and other perceptual characteristics (for reviews see Scharf 1970; Patterson and Moore 1986). Other terms for the critical band are "bark" (Zwicker and Fastl 1990) and the "equivalent rectangular bandwidth" or "ERB" (Patterson and Moore 1986).

4.2.1 Critical Ratio

The first measures of the critical band were obtained indirectly by Fletcher (1940). He assumed that when a tone is justmasked by wideband noise the energy in the critical band of the noise and the energy in the tone must be equal. Given this assumption, he estimated the critical bandwidth by subtracting the spectrum level of the noise (dB per Hz) from the level of the masked tone at threshold. He then determined the bandwidth of the noise (number of Hz) having energy equal to the energy in the tone. Repeating the measure with different spectrum level noises leads to approximately the same decibel difference between the threshold of the noise and the spectrum level of the tone. This pattern is expected if, as assumed, the critical band is independent of level. After more direct measures of the critical band were developed (see Section 4.2.2), this approximation of the critical bandwidth came to be known as the critical ratio (CR).

The CR is still measured because it minimizes the number of observations needed to estimate frequency resolution. If one has a background of white noise of known spectrum level, all one needs is to determine the threshold of a tone in that noise. Frequency resolution at a range of frequencies can be determined by measuring the masked threshold of several different frequency tones.

4.2.1.1 Comparison Between Frequency Discrimination and the Critical Ratio

In many species, the critical ratio is approximately 20 times larger than estimates of frequency discrimination (see Fig. 2.4), suggesting that both measures may depend on similar aspects of basilar membrane tuning (see Fay

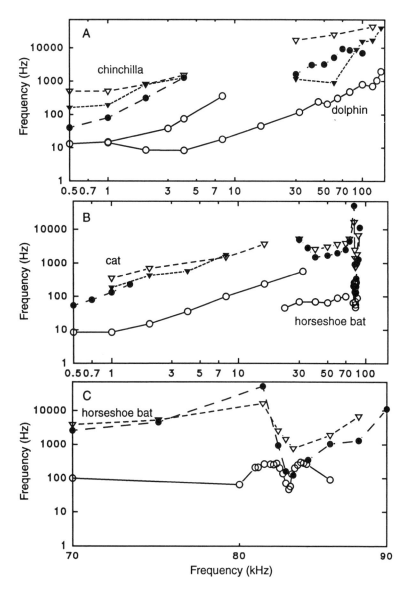

FIGURE 2.4. A comparison of frequency discrimination (dlf, open circles), critical ratio (CR, filled symbols), and critical band (CB, open triangles) measurements from four species. (A) Chinchilla *Chinchilla langier* pulsed-tone dlf (Nelson and Kiester 1978), CR and CB (Seaton and Trahoitis 1975). Atlantic bottlenose dolphin *Tursiops truncatus* FM dlf (Thompson and Herman 1975), CR (circles, Johnson 1968b; triangles, Au and Moore 1990), CB (Au and Moore 1990). (B) Cat *Felis catus* pulsed-tone dlf (Elliot, Stein, and Harrison 1960), CR and CB (Pickles 1975). Greater horseshoe bat *Rhinolophus ferrumequinum* pulsed-tone dlf (Heilmann-Rudolf 1984), CR (Long 1977), CB (Long 1980a,b). (C) The data from *Rhinolophus ferrumequinum* expanded to show the specialized region of the cochlea above 75 kHz.

1992). One species in which the distribution of frequency on the basilar membrane is non-uniform is the greater horseshoe bat (*Rhinolophus ferrumequinum*). As can be seen from Figure 2.4B, even though different experimental paradigms were used, estimates of frequency discrimination (Heilmann-Rudolf 1984) for tones below 70 kHz are $\frac{1}{20}$ of the CR estimates (Long 1977). This pattern does not hold in the region (above 75 kHz) in which frequency distribution on the cochlea is specialized (Vater 1988). Frequency discrimination and CR results at these frequencies are not the same function of frequency (see Fig. 2.4C).

The ease of measurement of the CR is offset by its disadvantages. These disadvantages are (a) the number of assumptions required (e.g., the assumption that the energy in the tone and the energy in the noise are equal at threshold is almost certainly invalid) and (b) the CR depends critically on the animal's performance efficiency as determined by training and measurement procedures. Very conservative estimates of threshold tend to overestimate the CR. Unfortunately, as discussed in Section 2.3, most psychoacoustic measures obtained from animals have been obtained using very conservative criteria.

4.2.2 Critical Band

Overestimation of the critical band due to conservative criteria can be avoided by using a direct measure of the critical band (CB). Different ways of obtaining direct estimates of the critical band are described below. These direct measures tend to be approximately 2.5 times larger than the CR, probably because the assumption that signal and noise energy are equal at threshold is not valid. Humans, dolphins, and the greater horseshoe bat (see Fig. 2.4) have smaller CRs and CBs than other animal species, consistent with the superior frequency discrimination performance in these species.

Estimates of the critical bandwidth increase with center frequency in a manner consistent with the distribution of frequency on the basilar membrane. One critical band appears to be a fixed distance on the basilar membrane (e.g., Greenwood 1990; Fay 1992). Cochlear specializations, such as those in *Rhinolophus ferrumequinum*, may disrupt this relationship (Long 1980a,b). As can been seen from Figure 2.4B, critical bands obtained from the greater horseshoe bat are 2.5 larger than CRs up to 75 kHz (i.e., in the frequency region where the cochlea appears unspecialized). However, the relationship doesn't hold in the region where the cochlea is specialized (Vater 1988). In this region (Fig. 4C), estimates of the CR, CB, and frequency discrimination all have different shapes.

The 2.5 ratio between CR and CB is not found in all nonhuman species (e.g., *Chinchilla langier*, Seaton and Trahiotis 1975; dolphin *Tursiops truncatus*, Au and Moore 1990; see Fig. 2.4A). In these species, the estimates of the critical ratio increase less with frequency than do the estimates of the critical band. It is difficult to tell if this discrepancy is due to biases stemming from experimental factors (such as changes in the spectrum of the noise in the ear) or to a real difference in the underlying anatomical and physiological processes. The discrepancy in the two measures was greatest in unpracticed animals, suggesting that it may stem from the subject's inability to attend to relevant cues. Other aspects of signal processing, such as the animal's ability to detect changes in intensity, may also influence estimates of the CR.

The major methods used to determine the critical band in research are outlined in Sections 4.2.2.1–4.2.2.4 and illustrated (using data from animal subjects) in Figure 2.5.

4.2.2.1 Band Narrowing

Thresholds are obtained for a tone centered in a wideband noise. The bandwidth of the noise is then narrowed. Estimates of the critical band from these data are based on the assumption that the critical band is a rectangular filter. If the total power of the noise is kept constant (see Fig. 2.5A) the threshold for the tone is expected to increase 3 dB every time the bandwidth is halved (i.e., the energy in the critical band would double). Once all the energy falls within a critical band, the threshold should remain constant.

An alternative approach is to keep the spectrum level of the noise constant and change the bandwidth by changing the filter setting (see Fig. 2.5B). In this more common procedure, halving the bandwidth of the wider bands of noise does not change the energy within the critical band and the threshold remains constant, even though the total energy is reduced 3 dB. However, once the noise is narrower than the critical band, the thresholds decrease 3 dB per octave. Unfortunately, the most important data points are obtained when the noise band is very narrow. Threshold estimates are very variable in narrowband noise (which has slow amplitude fluctuations and sounds similar to the signal; Bos and de Boer 1966). Traditionally, these data obtained using band-narrowing procedures were fit with two straight lines (0 dB slope and a 3 dB per octave slope) and the critical band was taken as the intersection of the lines.

4.2.2.2 Two-tone Masking of Narrowband Noise

In this procedure (Fig. 2.5C), the level of a narrowband noise centered between two tonal maskers with variable frequency separations is adjusted to determine the threshold (Zwicker 1954). This procedure is appealing because the results change rapidly at what is thought to be the edge of the critical band, giving a more repeatable estimate of the anticipated rectangular filter. As can be seen in Figure 2.5C, it also permits separate estimation of the upper and lower edges of the filter (e.g., Long [1980a,b] was able to determine that some of the critical bands in *Rhinolophus ferrumequinum* were very asymmetric). The sharp edges in the function may, however, be more a function of the generation of combination tones and other distortion products and not an independent characteristic of the filter (Greenwood 1991, 1992).

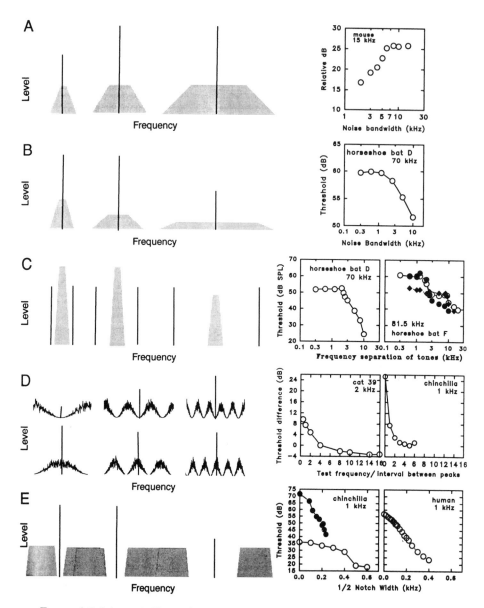

FIGURE 2.5. Schematic illustrations of the amplitude and frequency characteristics of the signal and maskers from different critical band paradigms are combined with representative data. (A) Band narrowing with constant spectrum level noise in the mouse *Mus musculus* (Ehret 1976a). (B) Band narrowing with constant power noise in the greater horseshoe bat *Rhinolophus ferrumequinum* (Long unpublished data). (C) Two-tone masking of noise in the greater horseshoe bat *Rhinolophus ferrumequinum* (Long 1980b). Closed symbols, single-tone maskers (triangles, low-frequency masker; circles, high-frequency masker). Open symbols, two-tone maskers. (D) Rippled noise in the cat *Felis catus* (Pickles 1979) and chinchilla *Chinchilla langier* (Niemiec, Yost, and Shofner 1992). (E) Band-stop noise in the chinchilla *Chinchilla langier*. Open symbols (forward masking, Halpern and Dallos 1986). Closed symbols (simultaneous masking, Niemiec, Yost, and Shofner 1992). Human *Homo sapiens* (Patterson 1976).

4.2.2.3 Rippled Noise

For a review of this procedure see Niemiec, Yost, and Shofner (1992). If noise is combined with a delayed version of itself, the result is periodic or "rippled" in the frequency domain. The frequency separation between the peaks and valleys of the ripple depend on the time delay. Thresholds are obtained with a tone at both a peak and a valley in the ripple of the noise, and the difference between these thresholds is used to determine the nature of the filter used in tone detection (Fig. 2.5D). The tone can be kept at a constant frequency and be placed at a peak or valley in the noise by adding or subtracting the delayed noise. When there is much less than one ripple cycle within a critical band, the amount of energy within the critical band will be large if the tone is at a peak and small if the tone is at a valley. When there are one or more ripples within a critical band, and if all of the energy within the critical band is summed, the total energy within a critical band should be independent of the position of the tone relative to a peak or a valley.

A predictable pattern of results (see Fig. 5D) with changes in the frequency separation of the ripples produces an estimate of the shape and bandwidth of the cochlear filter. This shape of the filter was initially thought to be approximately Gaussian (Houtgast 1977) and represented the beginning of a trend away from the concept of a rectangular filter. Symmetry in the shape of the auditory filter can be evaluated by comparing performance on the two half-power points (low frequency and high frequency between the minima and maxima in the noise). Representative data from the cat *Felis catus* (Pickles 1979) and chinchilla (Niemiec, Yost, and Shofner 1992) are compared with human data (Houtgast 1977) in Figure 2.5D. Pickles (1979) explained the negative threshold differences for closely spaced peaks (which are counter to theory) as evidence that the cat was detecting the tones in temporal fluctuations of the noise. Although Pickles (1979) estimated that the thresholds using rippled noise were consistent with those obtained using band-stop noise, Niemiec, Yost, and Shofner (1992) obtained estimates that were narrower than those obtained using band-stop noise.

4.2.2.4 Band-Stop or Notched Noise

Band-stop noise has become the most common procedure for measuring critical bands in humans since the 1970s. Recent research with animals have also employed this procedure (e.g., chinchilla, Halpern and Dallos 1986, Niemiec, Yost, and Shofner (1992); guinea pig *Cavia procellus*, Evans et al. 1992). In this paradigm, the tonal stimulus is usually centered between two bands of noise with very sharp edges (see Fig. 2.5E). The easiest way to generate noises with sharp cutoffs is to low-pass filter the noise and then multiply it with a tone. This produces a noise with twice the bandwidth of the low-pass noise and the rolloff remains the same in Hz (Patterson 1976). If the same noise is used for both the low- and high-frequency bands, the two noises will have the same envelope. The amplitude fluctuation of the noise will be greater than that of two independent noise sources because both noises will increase/decrease in level at the same time (Hall and Grose 1990). This could potentially reduce the threshold.

Placing bands of noise on either side of the signal reduces the subject's ability to attend a filter that is not centered on the stimulus (reviewed in Patterson and Moore 1986). If the two bands of noise are moved independently, one can also estimate the asymmetry of the critical band (reviewed in Patterson and Moore 1986). These detailed critical band measures have changed the concept of the rectangular critical band to a filter which has a rounded top and exponential slopes, abbreviated as ROEX for rounded exponential (reviewed in Patterson and Moore 1986). In addition, the band-stop noise procedure permits independent evaluation of filter bandwidth and processing efficiency (reviewed in Patterson and Moore 1986). The zero-notch width condition (i.e., no notch in the noise) is equivalent to the CR and is most influenced by processing efficiency and response bias (Glasberg et al. 1984). These authors thus recommended that the 0-Hz condition not be used in estimates of the equivalent rectangular bandwidth of the auditory filter (ERB).

Critical band estimates for the same subjects are narrower when the masker is band-stop noise than when it is rippled noise. This is true for humans (Glasberg et al. 1984), guinea pigs (Evans et al. 1992), and chinchillas (Niemiec, Yost, and Shofner 1992). Systematic differences between estimates of critical bands obtained with different procedures may be due to differences in filter bandwidth under different conditions or stem from the models used to extract a single estimate of bandwidth from the different sets of data.

4.2.3 Psychoacoustic and Neural Estimates of Frequency Resolution

When two signals are simultaneously presented to the cochlea, the response to each of the stimuli is modified (for a review see Delgutte 1990a). Physiological measures of frequency resolution (frequency tuning curves reviewed in Pickles 1986, 1988) are usually obtained with one signal (tone) presented at a time, and this is contrasted to the simultaneous presentation of signal and noise in most critical band experiments. In order to determine whether this simultaneous presentation of signal and masker modifies estimates of the critical band, experiments are done using nonsimultaneous presentations of masker and stimulus. The nonsimultaneous masking procedure used with animals is "forward masking" in which a long (usually 200 to 500 ms) masker precedes a brief signal (often 20 ms). The stimulus acting as the masker continues to elevate the threshold of a short signal even after termination of the stimulus (see Patterson and Moore 1986; Nelson et al. 1990). As predicted from neural measures of two-tone suppression, narrower estimates of the critical band are usually obtained when forward masking is used (reviewed in Pickles 1988).

Measures of basilar membrane motion and auditory nerve responses (reviewed in Patuzzi and Robertson 1988) are a nonlinear function of stimulus level. If one assumes that the critical band is dependent on distance along the basilar membrane, one must determine why the critical band is not equally dependent on the masker level. The use of a wideband signal, such as noise, may tend to linearize the cochlea and produce level-independent filters. Measures of filter width in the auditory nerve obtained with noise (reviewed in Møller 1977; Pickles 1986; Ruggero 1992) are also more stable across a range of levels. The narrower bandwidths obtained by forward masking in the cat are similar to the neural estimates of frequency resolution using noise stimuli obtained from the same species (Pickles 1980). In contrast, Evans et al. (1992) reported that psychoacoustic estimates of the critical band using band-stop and rippled noises in the guinea pig are similar to the bandwidths of auditory nerve fibers. For a review of the relationship between psychoacoustic measures of frequency resolution and physiological measures, see Pickles (1986).

Forward masking measures from humans are usually obtained using headphones. In this situation, there is no possibility that echoes from the masker will interfere with the detection of the probe. Unfortunately, much of the nonhuman data are obtained using speakers in environments in which echoes are probable. These echoes may modify the results obtained from both animals and humans (Halpern and Dallos 1986) to the extent that the results can no longer provide estimates of the auditory filter consistent with the rounded exponential shape characteristic of most human data. The detection of echoes in the forward masking estimates of Halpern and Dallos (1986) may explain why their estimates of the critical band are wider than the simultaneous masking estimates of Niemiec, Yost, and Shofner (1992) in Figure 2.5E.

Most communication sounds (reviewed in Gould 1983; Brown, Chapter 3), and other ethologically relevant stimuli, have bandwidths that are wider than a critical band. Consequently, the critical band provides a useful tool for understanding aspects of signal analysis. Zwicker and his colleagues (reviewed in Zwicker and Fastl 1990) have developed a more valid loudness meter for noise abatement research using the critical band or bark. This tool could be modified to assess the impact of noise and other complex stimuli on nonhuman species.

4.2.4 Integration Across Critical Bands

Past models of auditory processing have concentrated on explaining auditory perception by investigating processing within one critical band. Recent research (reviewed in Hall and Grose 1990; Moore, Glasberg, and Schooneveldt 1990) has established that there are many phenomena that cannot be explained unless one considers integration of information across critical bands. The random character of broadband noise used in most estimates of critical band and CR prevents the use of these cues because the lack of correlation across frequencies reduces the probability that the animal will integrate information across bands (Kidd et al. 1991). If the noise is not random, the animal may integrate information across critical bands. If additional noise bands with similar envelopes (correlated noise bands) are added to the original masker, there is less masking. Comodulation masking release (CMR) has been interpreted as a figure and ground phenomenon in which the figure (the signal) is separated from the masker (the background) because the figure has coherent amplitude or frequency fluctuations (reviewed in Hall and Grose 1990). Sounds in the environment are analyzed into an auditory scene. Different sounds are grouped together on the basis of cues that would suggest that different objects are generating the sounds (reviewed in Bregman 1990). Coherent amplitude and frequency modulations are one of the cues helping to determine which sounds are grouped together.

4.3 Masking with Tonal or Narrowband Stimuli

Psychoacoustic measures of frequency resolution obtained using tonal, or narrowband, signals and maskers are more similar to standard neurophysiological measures of single cell frequency selectivity (see Pickles 1988). One can either use a constant level tonal masker and vary the level of a tonal signal (masking functions) or determine the level of a tonal masker needed to stop detection of a low-sensation level (near threshold) tone. The former procedure produces masking functions that are qualitatively similar to the neural isointensity curves, while the second procedure gives psychoacoustic tuning curves that are similar to the neural isoresponse functions (reviewed in Pickles 1986, 1988; Ruggero 1992). Examples of psychoacoustic tuning curves obtained from the chinchilla (Ryan, Dallos, and McGee 1979; Salvi, Perry, and Hamernik 1982; Salvi et al. 1982a) are shown in Figure 2.6.

One common way to compare tuning curves is to estimate the Q or quality of the tuning curve. The frequency of the stimulus is divided by the bandwidth of the tuning curve at some constant level (usually 10 dB). In contrast to other measures of frequency resolution, there does not seem to be any trend for humans to have sharper tuning curves than those obtained from other animals.

The bandwidth of a neural tuning curve depends on the level at which the bandwidth is determined (Pickles 1988). Similarly, the bandwidth of the psychoacoustic tuning curve depends on the level of the stimulus (at both the sensation level and the sound pressure level) and the level at which the bandwidth is determined. The bandwidth of the psychoacoustic tuning curve in humans seems to depend largely on the level of the maskers needed to mask the probe (Nelson et al. 1990). If this is true for animals (and most results in humans are qualitatively similar to those for animals), cross-species comparison of psychoacoustic tuning curves may be contaminated by differences in the level of the signal that would determine the level of the maskers needed.

The similarity between neurophysiological tuning curves and psychoacoustic tuning curves may be misleading. The simultaneous presentation of masker and probe will induce suppression of one signal by the other (as in

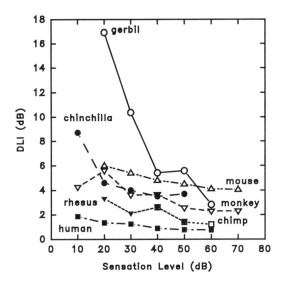

FIGURE 2.6. Intensity (level) discrimination thresholds (DLI) as a function of the level of the tone above threshold. Chinchilla *Chinchilla langier* (filled circles, Saunders, Shivapuja, and Salvi 1987). Rhesus monkey *Macaca mulata* (filled downward triangles, Sinnott, Petersen, and Hopp 1985). Mouse *Mus musculus* (open triangles, Ehret 1975). Chimpanzee *Pan trogodytes* (open squares, Kojima 1990). Gerbil *Meriones unguiculatus* (open circles), blue monkey *Cercopithecus mitus* (open downward triangles), and human *Homo sapiens* (filled squares) all in Sinnott, Brown, and Brown (1992).

measures of critical bands discussed in Section 4.2.2). Forward masking paradigms eliminate interactions between the masker and probe and usually produce curves that are more sharply tuned than the simultaneous masking tuning curves. As one would expect two-tone suppression to broaden the tuning curve, it has been hypothesized that the difference between the simultaneous masking tuning curve and the forward masking tuning curve stems from suppression.

Forward masking and simultaneous masking tuning curves may differ for reasons other than the existence of two-tone suppression (reviewed in O'Loughlin and Moore 1981; Moore and O'Loughlin 1986). Neural tuning curves measure the response at one place on the basilar membrane, but the behaving animals will respond to cues at any frequency (off-frequency listening). Thresholds obtained using simultaneous masking may reflect the subject's ability to detect interactions between the two tones, such as amplitude modulation due to beating between the masker and signal or the generation of combination tones. Narrow bands of noise have been used in simultaneous masking in an attempt to avoid the detection of combination tones. As Greenwood (1991, 1992) pointed out, this doesn't really remove the problem, it just makes it less obvious by replacing the narrow minima in the masking function caused by the detection of combination tones with a wider combination tone band. The detection of combination tones is related to frequency resolution and has been used to obtain an estimate of frequency resolution in the chinchilla (Long and Miller 1981). This estimate was very similar to other estimates of the critical band using wideband maskers in this species (reviewed in Long and Miller 1981) but wider than those determined by Niemiec, Yost, and Shofner (1992) using rippled and band-stop noises.

The subject in a forward masking experiment may have a problem determining whether signals close in frequency to the masker are additional tones or simply a continuation of the masker (reviewed in O'Loughlin and Moore 1981; Moore and O'Loughlin 1986). The failure to detect the probe may reflect a problem of temporal processing, instead of a failure of frequency resolution. The use of a narrow band of noise to replace either the probe (signal) or the masker reduces some of these problems by making the quality of the stimuli different. However, if the noise band used for the masker is narrow, the subject has a problem of detecting an additional fluctuation at the end of an already fluctuating masker (reviewed in O'Loughlin and Moore 1981; Moore and O'Loughlin 1986). Differences in the training and testing paradigms and interspecies differences may alter the cues attended to and produce significant differences in the sharpness of the tip of the tuning curves.

4.3.1 Neurophysiological Investigations of Tone-on-Tone Masking

Sinex and collaborators using chinchillas (Sinex and Harvey 1984, 1986; Mott, McDonald, and Sinex 1990) and Delgutte using the cat (1990b) have explored the physiological basis of tone-on-tone masking using stimulus presentation paradigms in physiological research that are similar to those used in psychophysical experiments. Identical stimuli were presented to a population of neurons with a range of different characteristic frequencies. All investigations established that the most robust response was often not from neurons near the frequency of the masker. This could provide a possible basis for off-frequency listening.

Delgutte (1990b) compared simultaneous and nonsimultaneous presentation of a constant level and frequency masker and a varying level and frequency probe in the cat. He determined that the shape of the masking function (which showed qualitative similarity to psychoacoustic masking functions) was due to both excitative and suppressive effects of the masker. Suppressive masking was greatest for frequencies well above the masker. He also suggested that the nonlinear character of the "upward spread of masking" was due to the nonlinear growth of suppression.

Neurophysiological models of psychoacoustic experiments can help the understanding of the cues available to the animal at the neuronal level. Investigation of populations of neurons can enhance this research by incorporating the animal's tendency to attend cues at any frequency. However, the information from single fibers cannot yet be used to determine how the whole animal will combine the information from different neurons. It may be possible to obtain some insight into how the animals integrate the information across neurons by conducting psychoacoustic and physiological research with the same experimental paradigms in the same species.

5. Intensity Processing

In addition to the frequency content of sound, the intensity of sound will convey useful information. The intensity difference limen (DLI) describes the minimum change in intensity needed for a subject to detect that a sound has changed in level can be investigated. As discussed in Section 2.4, the stimulus presentation paradigm influences the results. If procedures are compared within subjects (Viemeister and Bacon 1988; Turner, Zwislocki, and Filion 1989), the procedures usually used with human subjects require the largest changes in level for detection, placing human subjects potentially at a disadvantage. Human listeners can detect the smallest level changes, followed by primates and then rodents (reviewed in Sinott, Brown, and Brown 1992). The pattern of responses are, however, very similar in a range of species (see Fig. 2.7).

In contrast to the estimates of frequency discrimination, the pattern of results with changes in the frequency and level of the stimuli is independent of the experimental paradigm (e.g., Viemeister 1988a,b; Viemeister and Bacon 1988). The increment in level is a constant ratio of the two stimuli (a constant dB difference between the discriminable stimuli) when the stimuli or surrounding maskers are broadband (reviewed in Viemeister 1988a,b). Such stimuli prevent the recruitment of new cochlear regions when there is an increase in the stimulus level. When unmasked narrowband signals are discriminated, the increment needed is smaller than a constant ratio. If the data are plotted in terms of the difference in dB between the stimulus being compared, the size of the change needed for detection decreases when the level of the sound at a fixed frequency is increased (this is called the "near miss to Weber's Law"). The recruitment of additional cochlear regions (auditory nerve fibers) as the stimulus level increases is thought to be responsible for this phenomenon (reviewed in Viemeister 1988a,b). Intensity discrimination of high-frequency tones in human subjects does not follow this pattern. Discrimination is poorest at midsensation levels (levels not close to threshold but not more than 50 dB above threshold) under some measurement conditions (reviewed in Long and Cullen 1985). A similar pattern can be seen in some of the intensity discrimination measures from the mouse (Ehret 1975).

An apparent conflict exists between the narrow dynamic range of most auditory nerve fibers (see Pickles 1988; Ruggero 1992) and the ability of all animals to maintain excellent discrimination across a wide dynamic range (reviewed in Viemeister 1988; Winter and Palmer 1991). Recent neurophysiological models of intensity processing (reviewed in Winter and Palmer

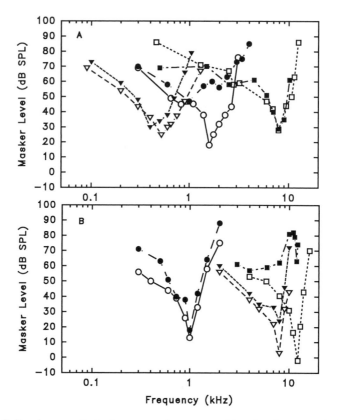

FIGURE 2.7. Psychoacoustic tuning curves from normal-hearing (open symbols) and hearing-impaired (filled symbols) chinchilla *Chinchilla langier*. (A) Probe (signal) at 500 Hz (triangles), 2 kHz (circles), and 8 kHz (squares) from normal-hearing chinchillas (Salvi et al. 1982a) and in chinchillas with a hearing impairment generated by a 5-day exposure to a 500-Hz octave band of noise at 95 dB SPL (Salvi, Perry, and Hamernik 1982). (B) Probe at 1 kHz (circles) and 8 kHz (triangles) before and after the generation of hearing impairment by kanamycin administration and a 11.9-kHz probe (open squares) after kanamycin (Ryan, Dallos, and McGee 1979). Probe at 12.25 kHz in a normal-hearing animal (filled squares, McGee, Ryan, and Dallos 1976).

1991) indicate that no such conflict exists. The information needed for intensity discrimination over a range of levels is available in one place on the cochlea, if one incorporates the recent evidence that the low spontaneous-rate fibers with high thresholds have a different pattern of saturation.

Differences in the level of two sequential tones will convey differences in distance to the sound source, a meaningful task for an animal trying to localize friend or foe. However, the more common task in nature would involve detection of differences in the levels of two or more simultaneously presented tones (e.g., judging which of several simultaneous sound sources is nearer or larger). Human research on "profile analysis" (reviewed in Green 1988; Kidd et al. 1991) has shown that comparing the level of simultaneously presented sounds is much easier than comparing the levels of sequentially presented tones.

6. Temporal Processing

Behaviorally meaningful stimuli usually fluctuate rapidly in amplitude. The animal's ability to process such stimuli depends on its ability to make use of temporal characteristics of the stimuli (such as changes in the envelope of the signal). In contrast to measures of frequency and intensity resolution (where most species show poorer frequency processing than humans), the animals appear to do as well as humans on measures of temporal processing (see Section 6). Models of frequency and intensity processing suggest that performance may depend on aspects of the basilar membrane traveling wave. This does not appear to be true for temporal resolution.

6.1 Temporal Integration

If the duration of a signal is less than about 250 ms, changing the duration of the sound affects the subject's ability to detect the sound (see Fig. 2.8A). The ear appears to act as an energy integrator. That is, as the duration doubles, the threshold declines 3 dB (reviewed in Fay 1992; Viemeister and Wakefield 1991). The duration over which integration occurs has been called the temporal integration constant. There is some, but not a strong, tendency for the duration of this constant to appear to be shorter at higher frequencies (Watson and Gengel 1969). Temporal integration characteristics are not dependent on a continuous presentation of the stimuli. Solecki and Gerken (1990) applied a model of temporal integration developed by Gerken, Bhat, and Hutchinson-Clutter (1990) that integrates temporal integration data from a wide range of species. This model assumes that threshold is a power function of duration and can be used to model data from contiguous and noncontiguous signals in humans and cats. One possible interpretation of temporal integration data is based on a neural counting model (see Fay 1992). Most temporal integration models fail for very short tones (less than 10 ms). The spectrum of such short tones has energy which spreads outside a single critical band and it appears that integration occurs within a critical band (reviewed in Moore 1989).

An alternative hypothesis for the results typically described by temporal integration is that the subject has more opportunities to sample a longer stimulus. A larger number of samples would then be available for later processing (Viemeister and Wakefield 1991). Such a multiple-looks hypothesis would allow selective integration of related information and permit exclusion of energy from different sources. The classification of sounds as coming from

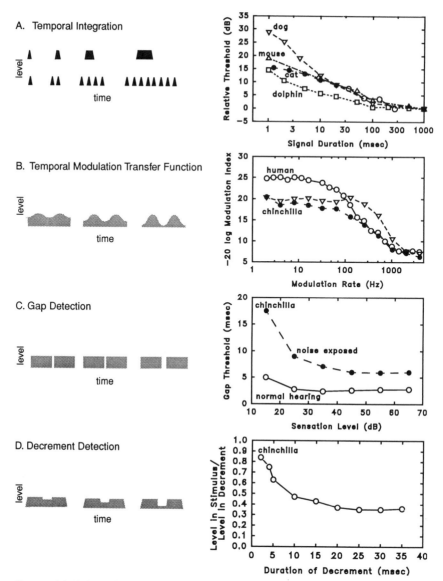

A. Temporal Integration

B. Temporal Modulation Transfer Function

C. Gap Detection

D. Decrement Detection

FIGURE 2.8. Schematic illustration of the patterns of temporal changes in the level of stimuli in experiments to evaluate temporal processing. (A) Temporal integration measures using continuous tones (open symbols) and pulsed tones (closed symbols). Cat *Felis catus* (6.25 kHz, circles, Solecki and Gerkin 1900). Dog *Canis canis* (1 kHz, downward triangles, Baru 1971). Mouse *Mus musculus* (40 kHz, triangles, Ehret 1976b). Bottlenose dolphin *Tursiops truncatus* (45 kHz, squares, Johnson 1968a). (B) Temporal modulation transfer function (the detection of sinusoidal modulation of noise). Human *Homo sapiens* (open circles, Viemeister 1979). Chinchilla *Chinchilla langier* normal hearing (open downward triangles) and after exposure to 4- and 8-kHz octave bands of noise (filled circles, Salvi et al. 1982b). (C) Gap detection in noise chinchilla *Chinchilla langier* before (open circles) and after exposure to a 4-kHz high-pass noise (filled circles, Salvi et al. 1982b). (D) The determination of the change in level needed to detect a fixed duration change in the chinchilla *Chinchilla langier* (Graf, Saunders, and Salvi 1992).

the same or different sources based on Gestalt principles (see Bregman 1990) would provide a basis for integrating relevant information (such as a tone) and ignoring noise inserted between the tones (Viemeister and Wakefield 1991).

6.2 Temporal Resolution

In experiments to test temporal resolution, animals are asked to detect changes in the envelope of sounds. The most common task requires the animal to detect the presence of a short period of silence embedded in noise (reviewed in Fay 1992). Noise is used as the stimulus because the spread of energy into other frequencies generated when a tone is turned on or off rapidly can generate spectral cues that may contaminate the results. The minimum gap needed for detection tends to be near 3 ms (see Fig. 2.8B). The length of the gap increases as one approaches threshold. The higher the frequencies in the noise and the wider the bandwidth of the noise, the shorter the gap needed (human data reviewed in Eddins, Hall, and Grose 1992). It has been hypothesized that the frequency dependence of gap detection is determined by the bandwidth of noise within a critical band (Shailer and Moore 1985). Eddins, Hall, and Grose (1992) suggested that the potential to use cues that occupy several critical bands may provide the basis for the improved performance with increases in the bandwidth of the noise.

Gap discrimination suffers from many of the same failings as the use of the CR. Estimates are very dependent on criteria, and one obtains no information about the animal's ability to detect the temporal fluctuations in environmentally meaningful stimuli. Temporal modulation transfer functions (TMTFs) provide one alternative approach (reviewed in Graf, Saunders, and Salvi 1992). The depth of modulation necessary for the animal to detect an envelope fluctuation is measured at a range of different modulation frequencies. The results are consistent with a low-pass filter characteristic (see Fig. 2.8B and C). The depth of modulation needed is constant up to 50–60 Hz (human) or 120 Hz (chinchilla), with a 3-dB slope after that (reviewed in Fay 1992). Noise is the preferred stimulus because the modulation of a pure tone produces sidebands in the spectrum of the sound, potentially converting the task into a frequency resolution task instead of a temporally based task (see Viemeister 1979). At high modulation rates, the overall level of the modulated and unmodulated stimuli differ and the task becomes an intensity discrimination task. Intensity discrimination also determines maximum performance at low fluctuation rates. Some of the differences in TMTF measures in humans and chinchillas tested with the same experimental paradigm (Salvi et al. 1982b) can be accounted for by differences in intensity discrimination performance (Graf, Saunders, and Salvi 1992).

More recent experiments with humans (e.g., Plack and Moore 1990, 1991) and chinchillas (Graf, Saunders, and Salvi 1992) have been model driven and have investigated the size of an increment (or decrement) in level of the stimulus needed at each duration for an animal to be able to detect a change in the signal (see Fig. 2.8D). If one assumes that there is a low-pass filter in the ear, one would expect that the shorter the duration, the larger the decrement in level needed for the detection of a change in the stimulus. If the decrement is very long, one would reach an asymptote that depended on the animal's intensity resolution capacity. By measuring the function of the size of the increment or decrement needed as the duration change is increased, it is possible to estimate a time constant. This approach gives consistent estimates of temporal resolution for temporal modulation transfer functions, increment detection, and gap detection (Forrest and Green 1987). The intensity decrement estimates of Graf, Saunders, and Salvi (1992) suggested that the human and chinchilla have the same integration time of 8 to 9 ms.

The combination of the multiple-looks theory of temporal integration proposed by Viemeister and Wakefield (1991) and the intensity decrement models discussed above permit the integration of the multiple temporal processing constants previously proposed (in Section 6.1 and 6.2) into a single time constant of approximately 8 ms.

7. Hearing Impairment

Comparative psychophysics provides a way of correlating changes in cochlear function due to noise exposure or ototoxic drugs with aspects of hearing impairment. Nonhuman animals can be used to evaluate the effects of ototoxic drugs or noise exposure on various psychoacoustic measures. At the end of the experiments, the animals are sacrificed and changes in the peripheral and central auditory system are correlated with changes in the psychoacoustic data. Prosen et al. (1990b) reviewed the effects of hair cell destruction on psychoacoustic thresholds while evaluating the effects of apical hair cell loss in chinchillas, guinea pigs, and monkeys. Damage to the outer hair cells leads to an elevation of thresholds (up to 40 dB) and a change in the frequency resolution of the cochlea (reviewed in Dallos 1988). Outer hair cells are currently believed to act as an energy source to amplify low-level sounds. Since this amplification is limited to a narrow range of frequencies, damage to the outer hair cells not only produces an increase in threshold but also leads to impaired frequency resolution and discrimination (reviewed in Dallos 1988). Inner hair cells are the transducers. In the absence of inner hair cells, the auditory nerve fibers from that region of the cochlea will not be stimulated and thresholds in that region are elevated.

Suprathreshold psychoacoustic processes have also been evaluated. Most of the research has concentrated on evaluating the impairment of frequency resolution associated with cochlear damage (reviewed in Clark and Bohne 1986; Smith, Moody, and Stebbins 1990). Many of the changes in psychophysical tuning curves following either noise exposure (Fig. 2.7A) or the use of ototoxic drugs (Fig. 2.7B) mirror the changes seen both in hearing-impaired human subjects (Tyler 1986) and in auditory nerve recordings (reviewed in Smith, Moody, and Stebbins 1990). Frequency resolution is impaired in all animals with hearing loss, and there can be impairment of frequency resolution at low frequencies, even when there is little or no hearing loss (reviewed in Prosen et al. 1990b). Changes in temporal processing in the chinchilla after noise exposure has been evaluated using both TMTF (Henderson et al. 1984) and gap detection (Salvi and Arehole 1985). Performance was poorer after noise exposure (see Fig. 2.8C and D). This change is probably due to the restriction of the range of frequencies available to the animal when the thresholds for high-frequency tones are poorer. The results are consistent with the human evidence that reducing the bandwidth of the noise (by filtering or hearing loss) reduces performance on temporal processing tasks (reviewed in Eddins, Hall & Grose 1992). Lesions of the apex of the cochlea (Prosen and Moody 1991) provide a cochlear model of individuals with low-frequency loss by establishing that much of their unexpectedly good low-frequency processing is based on responses of neurons at the base of the cochlea.

8. Otoacoustic Emissions

The development of research on otoacoustic emissions (reviewed by Probst, Lonsbury-Martin, and Martin 1991) provides a new tool for evaluating hearing in animals. Otoacoustic emissions are acoustic signals generated by the cochlea either in the absence of acoustic stimulation (spontaneous otoacoustic emissions) or in response to acoustic stimulation (evoked otoacoustic emissions). These sounds can be detected by placing a sensitive microphone in the ear canal and analyzing the resulting signal by computer or other signal analysis systems. This detection procedure is rapid and noninvasive and demands no active participation by the subject. Otoacoustic emissions, therefore, provide a common tool permitting comparison of psychoacoustic measures in humans and comparison with physiological measures in other species.

Otoacoustic emissions appear to be a side effect of the cochlear amplifier responsible for low thresholds and good frequency resolution (reviewed in Brownell 1990). Otoacoustic emissions can be used to provide information about cochlear function in the normal-hearing subject. For example, one can get estimates of the basilar membrane tuning in man by measuring the suppression of an emission by an external tone (reviewed in Long, Tubis, and Jones 1991). Similar results have been obtained in the chinchilla (Clark et al. 1984). The levels of external tones needed to suppress a spontaneous emission are very similar to the psychoacoustic tuning curves from the same subject in humans (Zurek 1981; Long 1984).

Spontaneous otoacoustic emissions are rarely measured in nonprimate animals (see review by Probst, Lonsbury-Martin, and Martin 1991), but recent reports from Japan indicate that they are not uncommon in young guinea pigs (Ohyama et al. 1991). Distortion product emissions (acoustic stimuli of predictable frequency generated when two tones are input to the cochlea) can be measured in all mammalian species tested to date. These include cat, chinchilla, gerbil, guinea pig, rabbit, mouse, monkey (reviewed in Probst Lonsbury-Martin and Martin 1991), and two species of bat, *Megaderma lyra and Carollia perspicillata* (Kössl 1992). Although there are differences in relative amplitude (the distortion product emissions in humans are considerably smaller than those in rodents), there are enough similarities (Brown and Gaskill 1990a) to make it a useful comparative tool. It is possible that distortion product otoacoustic emissions can be used to evaluate both sensitivity and basilar membrane tuning in humans (see review by Probst, Longsbury-Martin, and Martin 1991; Brown, Gaskill, and Williams 1992) and animal species (e.g., rabbit, Martin et al. 1988; guinea pig, Brown and Gaskill 1990b; bats, Kössl 1992). The greater frequency of spontaneous emissions in humans, together with the reduced level of distortion product emissions, may reflect differences in basilar membrane vibration and/or middle and inner ear impedance (Furst and Lapid 1988).

9. Summary

Obtaining psychoacoustic measures from animals is difficult and very time consuming. However, the time spent by many researchers has been invaluable in increasing the understanding of auditory processing. It has been established that the frequency range heard by an animal is determined primarily by the physics of the outer and middle ears. However, even animals that can hear very similar frequencies (detect that a sound is present) will process these sounds very differently. Animal psychoacoustics also helps the understanding of the differences in the reactions of different species to the same sounds.

The ability to tell that a sound has changed in frequency and the ability to resolve different frequency components of a complex sound (or background noise) are related and appear to depend on the length of the basilar membrane (provided that the membrane has no anatomical specializations). The only species that consistently do as well as human beings in frequency processing tasks are the echolocating animals. Humans are also better at detecting changes in the intensity of sounds than other animal species investigated. In contrast, measures of the animals' ability to detect changes in temporal aspects of the stimuli are comparable to those obtained from humans.

When combined with anatomical and neurophysiological research on the same species, psychoacoustics provides a basis for determining the peripheral processes underlying the perception of sounds by humans and animals. Research with animals that have been exposed to noise or drugs can help us understand the different patterns of hearing loss in the human population and potentlally aid in the development of better prostheses, such as the cochlear implant, for the hearing-impaired.

References

Au WWL, Moore PWB (1990) Critical ratio and critical bandwidth for the Atlantic bottlenose dolphin. J Acoust Soc Am 88:1635–1638.

Ayrapet'yants ESH, Konstantinov AI (1974) Echolocation in nature. An English translation of the National Technical Information Service, Springfield, VA, JPRS 63328-1 and -2.

Bargones JY, Marean DC, Werner LA (1992) Infant psychometric functions: Asymptotic performance. J Acoust Soc Am 91:2436 (Abstract).

Baru AV (1971) Behavioral thresholds and frequency difference limen as a function of sound duration in dogs deprived of the auditory cortex. In: Gersuni GV (ed) Sensory Processes at the Neuronal and Behavioral Levels. New York: Academic Press, pp. 757–763.

Bos CE, de Boer E (1966) Masking and discrimination. J Acoust Soc Am 39:708–715.

Bregman AS (1990) Auditory Scene Analysis: The Perceptual Organization of Sound. Cambridge, MA: MIT Press.

Brown AM, Gaskill SA (1990a) Measurement of acoustic distortion reveals underlying similarities between human and rodent mechanical responses. J Acoust Soc Am 88:840–849.

Brown AM, Gaskill SA (1990b) Can basilar membrane tuning be inferred from distortion measurement? In: Dallos P, Geisler CD, Matthews JW, Ruggero MA, Steele CR (eds) Lecture Notes in Biomathematics: The Mechanics and Biophysics of Hearing. Berlin: Springer-Verlag, pp. 164–169.

Brown AM, Gaskill SA, Williams DM (1992) Mechanical filtering of sound in the inner ear. Proc Soc Condon, Series Br 250:29–34

Brown CH, Waser PM (1984) Hearing and communication in blue monkeys (Cercopithecus mitis). Anim Behav 32:66–75.

Brownell WE (1990) Outer hair cell electromotility and otoacoustic emissions. Ear Hear 11:82–92.

Burdick CK (1979) The effect of behavioral paradigm on auditory dlscrimination learning: A literature review. J Aud Res 19 (Suppl 8): 59–82.

Burns EM, Viemeister NF (1981) Played again SAM: Further observations on the pitch of amplitude-modulated noise. J Acoust Soc Am 70:1655–1660.

Clark TD, Bohne BA (1986) Cochlear damage: Auditory correlates? In: Collins MJ, Glattke TJ, Harker LA (eds) Sensorineural Hearing Loss: Mechanisms, Diagnosis, and Treatment. Iowa City: University of Iowa Press, pp. 59–82.

Clark WW, Kim DO, Zurek PM, Bohne BA (1984) Spontaneous otoacoustic emissions in chinchilla ear canals: Correlation with histopathology and suppression by external tones. Hear Res 16:299–314.

Dallos P (1988) Cochlear neurobiology: Revolutionary development. American Speech-Language-Hearing Association 30:50–56.

Davey G (1989) Ecological Learning Theory. New York: Routledge.

Delgutte B (1990a) Two-tone rate suppression in auditory nerve fibers: Dependence on suppressor frequency and level. Hear Res 49:225–246.

Delgutte B (1990b) Physiological mechanisms of psychophysical masking observations from auditory nerve fibers. J Acoust Soc Am 87:791–809.

Dooling RJ. (1989) Perception of Complex, species-specific vocalizations by bird and humans. In: Dooling RJ, Hulse SH (eds) The Comparative Psychology of Audition: Perceiving Complex Sounds. Hillsdale, NJ: Lawrence Erlbaum Associates, Inc., pp. 423–444.

Eddins DA, Hall JW III, Grose JH (1992) The detection of temporal gaps as a function of frequency region and absolute noise bandwidth. J Acoust Soc Am 91:1069–1077.

Ehret G (1974) Age-dependent hearing loss in normal hearing mice. Naturwissenschaften 11:506.

Ehret G (1975) Frequency and intensity difference limens and nonlinearities in the ear of the house mouse (*Mus musculus*). J Comp Physiol 102:321–336.

Ehret G (1976a) Critical bands and filter characteristics of the ear of the house mouse (*Mus musculus*). Biol Cybernet 24:35–42.

Ehret G (1976b) Temporal auditory summation for pure tones and white noise in the house mouse (*Mus musculus*). J Acoust Soc Am 59:1421–1427.

Ehret G (1989) Hearing in the mouse. In: Dooling RJ, Hulse SH (eds) The Comparative Psychology of Audition: Perceiving Complex Sounds. Hillsdale, NJ: Lawrence Erlbaum Associates, pp. 3–32.

Elliot DN, Stein L, Harrison MJ (1960) Determination of absolute-intensity thresholds and frequency-difference thresholds in cats. J Acoust Soc Am 32:380–384.

Evans EF, Pratt SR, Spenner H, Cooper NP (1992) Comparisons of physiological and behavioral properties: Auditory frequency selectivity. In: Cazals Y, Horner K, Demany L (eds) Auditory Physiology and Perception. New York: Pergamon Press, pp. 159–169.

Fastl H (1978) Frequency discrimination for pulsed versus modulated tones. J Acoust Soc Am 63:275–277.

Fay RR (1974) Masking of tones by noise for the goldfish (*Carassius auratus*). J Comp Physiol Psychol 87:708–816.

Fay RR (1988) Hearing in Vertebrates: A Psychophysics Databook. Winnetka, IL: Hill-Fay Associates.

Fay RR (1992) Structure and function in sound discrimination among vertebrates. In: Webster DB, Fay RR, Popper AN (eds) The Evolutionary Biology of Hearing. New York: Springer-Verlag, pp. 229–263.

Fletcher H (1940) Auditory patterns. Rev Mod Phys 12:47–65.

Forrest TG, Green DM (1987) Detection of partially filled gaps in noise and the temporal modulation transfer function. J Acoust Soc Am 82:1933–1943.

Furst M, Lapid M (1988) A cochlear model for acoustic emissions. J Acoust Soc Am 84:215–221.

Gerken GM, Bhat VKH, Hutchison-Clutter M (1990) Auditory temporal integration and the power function model. J Acoust Soc Am 88:767–778.

Glasberg BR, Moore BCJ, Patterson RD, Nimmo-Smith I (1984) Dynamic range and asymmetry of the auditory filter. J Acoust Soc Am 76:419–427.

Gould E (1983) Mechanisms of mammalian auditory communication. In: Eisenberg JF, Kleiman DG (eds) Advances in the Study of Mammalian Behavior. The American Society of Mammalogists, Spec. Publ. #7, pp. 265–342.

Graf CJ, Saunders SS, Salvi RJ (1992) Detection of intensity decrements by the chinchilla. J Acoust Soc Am 91:1062–1068.

Green DM (1988) Profile Analysis: Auditory Intensity Discrimination. New York: Oxford University Press.

Greenwood DD (1990) A cochlear frequency-position function for several species—29 years later. J Acoust Soc Am 87:2592–2605.

Greenwood DD (1991) Critical bandwidth and consonance: Their operational definition in relation to cochlear nonlinearity and combination tones. Hear Res 54: 209–246.

Greenwood DD (1992) Erratum and comments re: Critical bandwidth and consonance: Their operational definitions in relation to cochlear nonlinearity and combination tones (Hear Res 54:209–246, 1991). Hear Res 59P:121–128.

Griffin DR (1971) The importance of atmospheric attenuation for the echolocation of bats (Chiroptera). Anim Behav 19:55–61.

Hall JW III, Grose JH (1990) Comodulation masking release and auditory grouping. J Acoust Soc Am 88:119–125.

Halpern DL, Dallos P (1986) Auditory filter shapes in the chinchilla. J Acoust Soc Am 80:765–775.

Harrison JM (1992) Avoiding conflicts between the natural behavior of the animal and the demands of discrimination experiments. J Acoust Soc Am 92:1331–1345.

Heffner RS, Heffner HE (1982) Hearing in the elephant (*Elephas maximus*): Absolute sensitivity, frequency discrimination, and sound localization. J Comp Psychol 96: 926–944.

Heffner RS, Heffner HE (1990) Vestigial hearing in a fossorial mammal, the pocket gopher (*Geomys bursarius*). Hear Res 46:239–252.

Heffner RS, Heffner HE (1991) Behavioral hearing range of the chinchilla. Hear Res 52:13–16.

Heffner RS, Heffner HE (1992) Evolution of sound localization in mammals. In: Webster DB, Fay RR, Popper AN (eds) The Evolutionary Biology of Hearing. New York: Springer-Verlag, pp. 691–715.

Heilmann-Rudolf U (1984) Das Frequenzunterscheidungsvermoegen bei der Grossen Hufeisennase. Dissertation der Eberhard-Karls-Universitat Tubingen, Germany.

Henderson D, Salvi RJ, Pavek G, Hamernik RP (1984) Amplitude modulation thresholds in chinchillas with high-frequency hearing loss. J Acoust Soc Am 75:1177–1183.

Hoffman HF, Ison JR (1992) Reflex modification and analysis of sensory processing in developmental and comparative research. In: Campbell BA, Hayne H, Richardson R (eds) Attention and Information Processing in Infants and Adults: Perspectives from Human and Animal Research. Hillsdale, NJ, Lawrence Erlbaum Associates, pp. 83–111.

Houtgast T (1977) Auditory-filter characteristics derived from direct-masking and pulsation-threshold data with a rippled-noise masker. J Acoust Soc Am 62:409–415.

Javel E, Mott JB (1988) Physiological and psychophysical correlates of temporal processes in hearing. Hear Res 34:275–294.

Jesteadt W, Sims SL (1975) Decision processes in frequency discrimination. J Acoust Soc Am 57:1161–1168.

Johnson CS (1968a) Relation between absolute threshold and duration-of-tone pulses in the botttlenosed porpoise. J Acoust Soc Am 43:757–763.

Johnson CS (1968b) Masked tonal thresholds in the bottlenosed porpoise. J Acoust Soc Am 44:965–967.

Ketten DR (1992) The marine mammal ear: Specializations for aquatic audition and echolocation. In: Webster DB, Fay RR, Popper AN (eds) The Evolutionary Biology of Hearing. New York: Springer-Verlag, pp. 717–750.

Kidd G Jr, Mason CR, Uchanski RM, Brantley MA (1991) Evaluation of simple models of auditory profile analysis using random reference spectra. J Acoust Soc Am 90:1340–1354.

Killion MC (1978) Revised estimate of minimum audible pressure: Where is the "missing 6 dB"? J Acoust Soc Am 63:1501–1508.

Kojima S (1990) Comparison of auditory functions in the chimpanzee and human. Folia Primatol 55:62–72.

Kössl M (1992) High frequency distortion products from the ears of two bat species, *Megaderma lyra* and *Carollia perspicillata*. Hear Res 60:156–164.

Kraus N, McGee T (1992) Electrophysiology of the human auditory system. In: Popper AN, Fay RR (eds) The Mammalian Auditory Pathway: Neurophysiology. New York: Springer-Verlag, pp. 335–403.

Long GR (1977) Masked auditory thresholds from the bat (*Rhinolophus ferrumequinum*). J Comp Physiol 116:247–255.

Long GR (1980a) Further studies of masking in the greater horseshoe bat (*Rhinolophus ferrumequinum*). In: Busnel RG, Fish JF (eds) Animal Sonar Systems. New York: Plenum Press, pp. 929–932.

Long GR (1980b) Some psychophysical measurements of frequency processing in the greater horseshoe bat. In: van den Brink G, Bilsen FA (eds) Psychophysical, Physiological and Behavioural Studies in Hearing. Delft, The Netherlands: Delft University Press, pp. 132–135.

Long GR (1983) Psychoacoustical measures of frequency processing in mammals. In: Fay RR, Gourevitch G (eds) Hearing and Other Senses: Papers in Honor of E. G. Wever. Groton, CT: Amphora Press, pp. 230–246.

Long GR (1984) The microstructure of quiet and masked thresholds. Hear Res 15:73–87.

Long GR, Clark WW (1984) Detection of frequency and rate modulation by the chinchilla. J Acoust Soc Am 75:1184–1190.

Long GR, Cullen JK Jr (1985) Intensity limens at high frequencies. J Acoust Soc Am 78:507–513.

Long GR, Miller JD (1981) Tone-on-tone masking in the chinchilla. Hear Res 4:279–285.

Long GR, Schnitzler HU (1975) Behavioral audiograms from the bat (*Rhinolophus ferrumequinum*). J Comp Physiol 100:211–219.

Long GR, Tubis A, Jones KL (1991) Modeling synchronization and suppression of spontaneous otoacoustic emissions using van der Pol oscillators: Effects of aspirin administration. J Acoust Soc Am 89:1201–1212.

Mackintosh NJ (1974) The Psychology of Animal Learning. New York: Academic Press.

MacMillan NA, Creelman CD (1991) Detection Theory: A User's Guide. Cambridge: Cambridge University Press.

McGee T, Ryan A, Dallos P (1976) Psychophysical tuning curves of chinchillas. J Acoust Soc Am 60:1146–1150.

Martin GK, Stagner BB, Coats AC, Lonsbury-Martin BL (1988) Endolymphatic hydrops in rabbits: Behavioral thresholds, acoustic distortion products, and cochlear pathology. In: Nadol JB Jr (ed) Second International Symposium on Meniere's Disease: Pathogenesis, Pathophysiology, Diagnosis and Treatment. Boston, MA: Harvard University Press, pp. 205–219.

Michelsen A (1992) Hearing and sound communication in small animals: Evolutionary adaptations to the laws of physics. In: Webster DB, Fay RR, Popper AN (eds), The Evolutionary Biology of Hearing. New York: Springer-Verlag, pp. 61–77.

Møller AR (1977) Frequency selectivity of single auditory nerve fibers in response to broadband noise stimuli. J Acoust Soc Am 62:135–142.

Moody DB (1970) Reaction time as an index of sensory function. In: Stebbins WC (ed) Animal Psychophysics: The Design and Conduct of Sensory Experiments. New York: Appleton-Century-Crofts, pp. 277–301.

Moore BCJ (1986) Frequency Selectivity in Hearing. London: Academic Press.

Moore BCJ (1989) An Introduction to the Psychology of Hearing. San Diego, CA: Academic Press.

Moore BCJ, O'Loughlin BJ (1986) The use of nonsimultaneous masking to measure frequency selectivity and suppression. In: Moore BCJ (ed) Frequency Selectivity in Hearing. London: Academic Press, pp. 179–250.

Moore BCJ, Glasberg BR, Schooneveldt GP (1990) Across-channel masking and comodulation masking release. J Acoust Soc Am 87:1683–1694.

Moore BCJ, Glasberg BR, Simpson A (1992) Evaluation of a method of simulating reduced frequency selectivity. J Acoust Soc Am 91:3402–3423.

Moran G (1975) Severe food deprivation: Some thoughts regarding its exclusive use. Psychol Bull 82:543–557.

Mott JB, McDonald LP, Sinex DG (1990) Neural correlates of psychophysical release from masking. J Acoust Soc Am 88:2682–2691.

Nelson DA, Kiester TE (1978) Frequency discrimination in the chinchilla. J Acoust Soc Am 64:114–126.

Nelson DA, Chargo SJ, Kopun JG, Freyman RL (1990) Effects of stimulus level on forward-masked psychophysical tuning curves in quiet and in noise. J Acoust Soc Am 88:2143–2151.

Niemiec AJ, Yost WA, Shofner WP (1992) Behavioral measures of frequency selectivity in the chinchilla. J Acoust Soc Am 92:2636–2649.

Ohyama K, Wada H, Kobayashi T, Takasaka T (1991) Spontaneous otoacoustic emissions in the guinea pig. Hear Res 56:111–121.

O'Loughlin BJ, Moore BCJ (1981) Off-frequency listening: Effects on psychoacoustical tuning curves obtained in simultaneous and forward masking. J Acoust Soc Am 69:1119–1125.

Patterson RD (1976) Auditory filter shapes derived with noise stimuli. J Acoust Soc Am 59:640–654.

Patterson RD, Moore BCJ (1986) Auditory filters and excitation patterns as representations of frequency resolution. In: Moore BCJ (ed) Frequency Selectivity in Hearing. London: Academic Press, pp. 123–177.

Patuzzi R, Robertson D (1988) Tuning in the mammalian cochlea. Physiol Rev 68: 1009–1082.

Payne KB, Langbauer WR Jr, Thomas EM (1986) Infrasonic calls of the Asian elephant (Elephas maximus). Behav Ecol Sociobiol 18:297–301.

Pfingst BE (1988) Comparison of psychophysical and neurophysiological studies of cochlear implants. Hear Res 34: 243–252.

Pfingst BE, Rai DT (1990) Effects of level on nonspectral frequency difference limens for electrical and acoustic stimuli. Hear Res 50:43–56.

Pfingst BE, Rush NL (1987) Discrimination of simultaneous frequency and level changes in electrical stimuli. Ann Otol Rhinol Laryngol 96(Suppl 128):34–37.

Pickles JO (1975) Normal critical bands in the cat. Acta Otolaryngol 80:245–254.

Pickles JO (1979) Psychophysical frequency resolution in the cat as determined by simultaneous masking and its relation to auditory nerve resolution. J Acoust Soc Am 66:1725–1732.

Pickles JO (1980) Psychophysical frequency resolution in the cat studied with forward masking. In: van den Brink G, Bilsen FA (eds) Psychophysical, Physiological and Behavioural Studies in Hearing. Delft, The Netherlands: Delft University Press, pp. 118–126.

Pickles JO (1986) The neurophysiological basis of frequency selectivity. In: Moore BCJ (ed) Frequency Selectivity in Hearing. London: Academic Press, pp. 51–121.

Pickles JO (1988) An Introduction to the Physiology of Hearing. San Diego, CA: Academlc Press.

Plack CJ, Moore BCJ (1990) Temporal window shape as a function of frequency and level. J Acoust Soc Am 87:2178–2187.

Plack CJ, Moore BCJ (1991) Decrement detection in normal and impaired ears. J Acoust Soc Am 90:3069–3076.

Pollak GD, Casseday JH (1989) The Neural Basis of Echolocation in Bats. Berlin: Springer-Verlag.

Poole JH, Payne KB, Langbauer WR Jr, Moss CJ (1988) The social contexts of some very low frequency calls of African elephants. Behav Ecol Sociobiol 22:385–392.

Probst R, Longsbury-Martin BL, Martin GK (1991) A review of otoacoustic emissions. J Acoust Soc Am 89:2027–2067.

Prosen CA, Moody DB (1991) Low-frequency detection and discrimination following apical hair cell destruction. Hear Res 57:142–152.

Prosen CA, Halpern DL, Dallos P (1989) Frequency difference limens in normal and sensorineural hearing impaired chinchillas. J Acoust Soc Am 85:1302–1313.

Prosen CA, Moody DB, Sommers MS, Stebbins WC (1990a) Frequency discrimination in the monkey. J Acoust Soc Am 88:2152–2158.

Prosen CA, Moody DB, Stebbins WC, Smith DW, Sommers MS, Brown JN, Altschuler RA, Hawkins JE Jr (1990b) Apical hair cells and hearing. Hear Res 44:179–194.

Rado R, Himelfarb M, Arensburg B, Terkel J, Wollberg Z (1989) Are seismic communication signals transmitted by bone conduction in the blind mole rat? Hear Res 41:23–30.

Ravicz ME, Rosowski JJ, Voigt HF (1992) Sound-power collection by the auditory periphery of the Mongolian gerbil *Meriones unguiculatus*. I. Middle ear input impedance. J Acoust Soc Am 92:157–177.

Relkin EM (1988) Introduction to the analysis of middle ear function. In: Jahn AF, Santos-Sacchi J (eds) Physiology of the Ear. New York: Raven Press, pp. 103–123.

Rice JJ, May BJ, Spirou GA, Young ED (1992) Pinna-based spectral cues for sound localization in cat. Hear Res 58:132–152.

Rosowski JJ (1992) Hearing in transitional mammals: Predictions from the middle ear anatomy and hearing capabilities of extant animals. In: Webster DB, Fay RR, Popper AN (eds) The Evolutionary Biology of Hearing. New York: Springer-Verlag, pp. 615–631.

Ruggero MA (1992) Physiology and coding of sound in the auditory nerve. In: Popper AN, Fay RR (eds) The Mammalian Auditory Pathway: Neurophysiology. New York: Springer-Verlag, pp. 34–93.

Ryan A (1976) Hearing sensitivity of the mongolian gerbil, *Meriones unguiculatus.* J Acoust Soc Am 59:1222–1226.

Ryan A, Dallos P, McGee T (1979) Psychophysical tuning curves and auditory thresholds after hair cell damage in the chinchilla. J Acoust Soc Am 66:370–378.

Salvi RJ, Arehole S (1985) Gap detection in chinchillas with temporary high-frequency hearing loss. J Acoust Soc Am 77:1173–1177.

Salvi RJ, Perry JW, Hamernik RP (1982) Relationships between cochlear pathologies and auditory nerve and behavioral responses following acoustic trauma. In: Hamernik RP, Henderson D, Salvi RJ (eds) New Perspectives on Noise-Induced Hearing Loss. New York: Raven Press, pp. 165–188.

Salvi RJ, Ahroon WA, Perry JW, Gunnarson AD, Henderson D (1982a) Comparison of psychophysical and evoked-potential tuning curves in the chinchilla. Am J Otolaryngol 3:408–416.

Salvi RJ, Giraudi DM, Henderson D, Hamernik RP (1982b) Detection of sinusoidal amplitude-modulated noise by the chinchilla. J Acoust Soc Am 71:424–429.

Saunders SS, Shivapuja BG, Salvi RJ (1987) Auditory intensity discrimination in the chinchilla. J Acoust Soc Am 82:1604–1607.

Scharf B (1970) Critical bands. In: Tobias JV (ed) Foundations of Modern Auditory Theory, Volume 1. New York: Academic Press, pp. 159–202.

Seaton WH, Trahiotis C (1975) Comparison of critical ratios and critical bands in the monaural chinchilla. J Acoust Soc Am 57:193–199.

Shailer MJ, Moore BCJ (1985) Detection of temporal gaps in bandlimited noise: Effects of variations in bandwidth and signal-to-masker ratio. J Acoust Soc Am 77:635–639.

Sinex DG, Havey DC (1984) Correlates of tone-on-tone masked thresholds in the chinchilla auditory nerve. Hear Res 13:285–292.

Sinex DG, Havey DC (1986) Neural mechanisms of tone-on-tone masking: Patterns of discharge rate and discharge synchrony related to rates of spontaneous discharge in chinchilla auditory nerve. J Neurophysiol 56:1763–1780.

Sinnott JM, Petersen M, Hopp S (1985) Frequency and intensity discrimination in humans and monkeys. J Acoust Soc Am 78:1977–1985.

Sinnott JM, Brown CH, Brown FE (1992) Frequency and intensity discrimination in Mongolian gerbils, African monkeys and humans. Hear Res 59:205–212.

Sivian LJ, White SD (1933) On minimum audible sound fields. J Acoust Soc Am 4:288–321.

Smith DW, Moody DB, Stebbins WC (1990) Auditory frequency selectivity. In: Berkeley MA, Stebbins WC (eds) Comparative Perception, Volume 1. New York: John Wiley and Sons, pp. 67–95.

Solecki JM, Gerken GM (1990) Auditory temporal integration in the normal hearing and hearing-impaired cat. J Acoust Soc Am 88:779–785.

Stebbins WC (1973) Hearing of Old World monkeys (*Cercopithecinae*). Am J of Phys Anthropol 38:357–364.

Stebbins WC (1990) Perception in animal behavior. In: Berkeley MA, Stebbins WC (eds) Comparative Perception, Volume 1. New York: John Wiley and Sons, pp. 1–26.

Stinson MR, Lawton BW (1989) Specification of the geometry of the human ear canal for the prediction of sound-pressure level distribution. J Acoust Soc Am 85:2492–2503.

Thompson RK, Herman LM (1975) Underwater frequency discrimination in the bottlenose dolphin (1–140 kHz) and the human (1–8 kHz). J Acoust Soc Am 57:943–948.

Turner C, Zwislocki J, Filion P (1989) Intensity discrimination determined with two paradigms in normal and hearing-impaired subjects. J Acoust Soc Am 86:109–115.

Tyler RS (1986) Frequency resolution in hearing-impaired listeners. In: Moore BCJ (ed) Frequency Selectivity in Hearing. London: Academic Press, pp. 309–371.

Vater M (1988) Cochlear physiology and anatomy in bats. In: Nachtigall PE, Moore PWB (eds) Animal Sonar, Processes and Performance. New York: Plenum Press, pp. 225–243.

Viemeister NF (1979) Temporal modulation transfer functions based on modulation thresholds. J Acoust Soc Am 66:1364–1380.

Viemeister NF (1988a) Psychophysical aspects of auditory intensity coding. In: Edelman GM, Gall WE, Cowan WM (eds) Auditory Function: Neurobiological Bases of Hearing. New York: John Wiley and Sons, pp. 213–241.

Viemeister NF (1988b) Intensity coding and the dynamic range problem. Hear Res 34. 267–274.

Viemeister NF, Bacon SP (1988) Intensity discrimination, increment detection, and magnitude estimation for 1-kHz tones. J Acoust Soc Am 84:172–178.

Viemeister NF, Schlauch RS (1992) Issues in infant psychoacoustics. In: Werner LA, Rubel EW (eds) Developmental Psychoacoustics. Washington, DC: American Psychological Association, 191–209.

Viemeister NF, Wakefield GH (1991) Temporal integration and multiple looks. J Acoust Soc Am 90:858–865.

Watson CS (1963) Masking of tones by noise for the cat. J Acoust Soc Am 35:167–172.

Watson CS, Gengel RW (1969) Signal duration and signal frequency in relation to auditory sensitivity. J Acoust Soc Am 46:989–997.

Webster DB, Plassman W (1992) Parallel evolution of low-frequency sensitivity in Old World and New World desert rodents. In: Webster DB, Fay RR, Popper AN (eds) The Evolutionary Biology of Hearing. New York: Springer-Verlag, pp. 633–636.

Wier CC, Jesteadt W, Green DM (1976) Frequency discrimination as a function of frequency and sensation level. J Acoust Soc Am 61:178–184.

Wightman F, Allen P (1992) Individual differences in auditory capability among preschool children. In: Werner LA, Rubel EW (eds) Developmental Psychoacoustics. Washington, DC: American Psychological Association, 113–133.

Wiley RH, Richards DG (1978) Physical constraints on acoustic communication in the atmosphere: Implications for the evolution of animal vocalizations. Behav Ecol Sociobiol 3:69–94.

Winter IM, Palmer AR (1991) Intensity coding in low-frequency auditory nerve fibers of the guinea pig. J Acoust Soc Am 90:1958–1967.

Zurek PM (1981) Spontaneous narrowband acoustic signals emitted by human ears. J Acoust Soc Am 69:514–523.

Zwicker E (1954) Die Verdeckung von Schmalbandgerauschen durch Sinustone. Acoustica 4:415–420.

Zwicker E, Fastl H (1990) Psychoacoustics: Facts and Models. Berlin, Heidelberg: Springer-Verlag.

3
Sound Localization

CHARLES H. BROWN

1. Introduction

The ongoing acoustic ambience of all terrestrial habitats is a reflection of the sum of the biotic sources of sound in the environment (e.g., vocalizations, insect wing noise, and the staccato hammering of woodpeckers) and the non-biotic sources of sound (e.g., wind- and rain-induced vegetation movement and thunder). Different habitats are apt to sound distinctive, and perhaps the initial constraint on the sounds is rendered by the constituents of the soil (Linskens et al. 1976). Soil differences, variations in the topography of the landscape (its contour, elevation, humidity, and temperature), and the way in which wind and water move across the land render each portion of the environment more or less favorable for habitation by particular plant and animal species. Consequently, the distribution and density of both the flora and fauna varies across the land, and, in turn, these variations establish the characteristics of the "acoustic landscape." In this way, each place in nature comes to have its own special sound (Gish and Morton 1981; Brenowitz 1982; Waser and Brown 1984, 1986; Brown and Waser 1988). Within the constraints imposed by the properties of each acoustic landscape, hearing and communication systems evolved, and for any sound in nature, two fundamental questions are the most critical for perception: where is the source of the sound (sound localization analysis), and what is the source of the sound (auditory image analysis)?

In natural environments, the approach of a competitor, a predator, a relative, a mate, or one's prey may be conveyed by subtle fluctuations within the acoustic background. In many instances, it is likely that the early detection of an intruder is conveyed not by a sound which is unusual or uncommon because of its amplitude or frequency composition, but rather by a sound which is distinctive chiefly because it occurred at an "inappropriate" location within the acoustic landscape. Here, one's ability to survive depends not on unusual sound detection capabilities, but rather on a sound localization system that permits a listener to effortlessly, yet vigilantly, track the relative positions of the sources of sounds that signal safety or danger. Moreover, the absence of a "safe" sound may be as significant to many birds and mammals as is the presence of an "unsafe" one, for an intruder's approach may be disclosed by either the production of unexpected sounds or the abrupt cessation of "expected" sounds that were previously sustained or ongoing in some regular pattern. Movements made unstealthily will disrupt the chorus of cicadas or the sounds of birds or other animals, and a ripple of silence may spread across the landscape signaling that something (or someone) is nearby. The most subtle acoustic changes may be biologically the most telling. Clumsy predators are apt to go hungry, and an evolutionary premium has been placed upon the ability of most animals to quickly discern the position of a sound that does not belong (or the position of an unexpected cessation of those sounds that do belong). In the struggle for survival, the determination of the origin of a signal may be assigned a level of importance that equals or exceeds that of being able to recognize the sound or being able to identify the perpetrator of the disturbance. It is in this biological context that the mechanisms underlying sound localization evolved, and through the course of the succession of animals on earth, the sound localization abilities of many species have come to exhibit remarkable acuity.

The position of the source is a cardinal perceptual attribute of sound. Under normal conditions, the source of a stimulus is instantly and effortlessly assigned a position with reference to the orientation of the listener. The localization of sound is seemingly reflexive, the perception of direction is "instantaneous," and localization does not appear to be derived by some kind of deductive cognitive process. That is, under most conditions, listeners do not actively think about having to triangulate the possible origin of the sound given what they heard at their two ears. Just as a sound is perceived as having some magnitude or pitch (or noisiness), loudness, and duration, it is also perceived as having a distance dimension (it is near or far), an elevation dimension (above or below), and an azimuth dimension (left or right of the observer). Only when human subjects wear earphones do sounds routinely lack a coherent spatial image but, under these conditions, the normal filtering characteristics of the external ear and ear canal are bypassed (see Rosowski, Chapter 6) and the normal correlation between the timing and amplitude of the signals at the two ears is violated.

Batteau et al. (1965) noted that many sounds presented through earphones are reported to have an origin somewhere inside the listener's head. They showed that sounds presented through earphones would be perceived as having a normal external position and could be accurately located in space if the signals fed to the left ear and right ear originated from microphones positioned approximately 17.5 cm apart (a normal head width) and if the microphones were fitted with replicas of human pinnae. The apparent origin of the signal is "external" to the listener when sounds are presented this way, and if the position of a sound source delivered to the microphone array is moved to the left or to the right, the perceived location of the sound source moves accordingly. If the artificial pinnae are removed from the microphones, or if the normal free-field–to–eardrum transfer functions are artificially manipulated, localization accuracy suffers (Batteau et al. 1965; Wightman and Kistler 1989a,b, 1992).

It is known that the position of sound is a cardinal quality of auditory perception in adult human subjects, and there is good reason to believe that the same is true for human infants and for most vertebrates. That is to say, the position of a startling sound appears to "command" most vertebrates to orient towards its site of origin. Though auditory experience may modify and adjust localization during development (Knudsen 1983; Knudsen, Knudsen, and Esterly 1984), reflexive orientation to sound position is evident at or near birth in a wide variety of subjects including laughing gulls, *Larus atricilla* (Beer 1969, 1970), Peking ducklings, *Anas platyrhynchos* (Gottlieb 1965), infant cats (Clements and Kelly 1978a), rats (Potash and Kelly 1980), guinea pigs (Clements and Kelly 1978b), and humans (Wertheimer 1961; Muir and Field 1979). The data suggests that most vertebrates, including both altricial and precocial species, are able to reflexively locate the origin of sound almost as soon as the ear canal opens and they are able to hear. It is possible for many species that the perceptual ability to localize sound may be intact prior to birth, and its expression may be obscured until the organism has matured to the point at which it has gained the strength necessary to orient the head.

In many organisms, sound localization mechanisms may initiate and actively guide saccadic eye movements to the site of potentially important events. Animals with binocular frontal visual systems, such as the advanced primates, have limited peripheral or hemispheric vision, and these species may be particularly dependent upon a high-acuity directional hearing system to rapidly direct the eyes to the location of a disturbance (Harrison and Irving 1966). Furthermore, R. S. Heffner and H. E. Heffner (1985) have observed that the more restricted the width of the horizontal binocular visual field in various mammals, the greater the acuity of their sound localization abilities. This enhanced acuity may be critical for accurately aiming the eyes.

The perception of many events is bimodal. Speech perception, for example, is influenced by both visual information regarding tongue and lip configuration and the corresponding acoustic signal. When these two modalities of information are out of synchrony, or artificially separated in space, the result is discomforting to both adult and infant human subjects (Aronson and Rosenbloom 1971; Mendelson and Haith 1976). The preservation of the normal congruence between visual and auditory space is important for the development of sound location discriminations in animals (Beecher and Harrison 1971). Animals appear to be prepared to learn to direct visually guided responses towards objects positioned at the origin of sounds and correspondingly unprepared to learn to direct responses towards objects which have been repositioned so that the contiguity of visual and auditory space has been violated (Beecher and Harrison 1971; Harrison et al. 1971; Harrison 1990, 1992). For an organism to be able to react appropriately to events occurring at different locations in space, it is necessary that the visual and acoustic perceptual maps be aligned and in register. Abnormal visual experience early in development alters sound localization in rats (Spigelman and Bryden 1967), ferrets, *Mustela putorious* (King et al. 1988), and barn owls, *Tyto alba* (Knudsen and Knudsen 1985, 1990; Knudsen and Brainard 1991; Knudsen, Esterly, and du Lac 1991). The data suggests that visual experience early during development helps calibrate midbrain structures involved in spatial perception (the mammalian superior colliculus or the avian homologue, the optic tectum) to auditory signals arising from different spatial locations. As the interocular and interaural distances expand with growth of the head, it is likely that recalibration of the nervous system is necessary to maintain spatial registration between both modalities. These observations emphasize the fact that perception is frequently bimodal and that sound localization often plays an important role in shifting the site of gaze.

Though sound localization mechanisms evolved because of their significance to survival in the natural world (Masterton, Heffner, and Ravizza 1969), sound localization abilities have nearly always been studied in synthetic, quiet, echo-free environments (or even with earphones), and testing has often been conducted with tones, clicks, or bandlimited bursts of noise. The intent of this tradition has been to assess the absolute limits of precision of directional hearing, though at the expense of exploring how well sound localization abilities function under more normal conditions. In the sections that follow, I will discuss the physical cues for sound localization available to terrestrial vertebrates and the behavioral methodologies commonly used to assess the sound localization capabilities of animals, and then I will survey the sound localization abilities of selected mammals.

2. Localization Cues

The origin of sound in space is referenced relative to the orientation of the listener, and sound position is accordingly expressed in terms of its azimuth, elevation, and distance from the listener. All terrestrial vertebrates are sensitive to sound pressure, and the location of a source of sound is derived by the nervous system through the comparison of differences in the pressure wave incident at each ear and through the comparison of transformations in sound "quality" induced by the habitat and the shape of the head and pinna.

2.1 Sound Source Azimuth: The Horizontal Coordinate Of Sound Location

Sound localization is dependent upon the comparison of the sound waves incident at each ear in most terrestrial vertebrates. These interaural (or binaural) differences are the result of two factors: (1) the difference in distance (Δd) the sound wave must travel to reach the tympanic membrane of the two ears, and (2) differences in the level (or amplitude) of the signal incident at each ear. The first factor results in differences in the *time-of-arrival* and in differences in the *phase* of the signal at each ear. When a sound is presented from a position off to one side of a listener (not at the midline, or 0° azimuth), corresponding points in the sound wave will necessarily be received by the "near" ear (the ear on the side of the head which is toward the source of the sound) before it reaches the "far" ear. The velocity of sound in air is nominally 343 meters per second; given this velocity, for each additional centimeter the sound wave must travel to reach the far ear, the wave will arrive 29 μs later than it will at the near ear. Hence, interaural differences in the *time-of-arrival* of corresponding points in the sound wave may serve as one of the principle cues for directional hearing.

In the case where the stimulus is a simple sustained cyclic wave, such as a pure tone, differences in the arrival time of the near- and far-ear wave-forms will result in interaural differences in the phase of the wave as long as the arrival time difference is not equal to the period (or integral multiples of the period) of the signal. For example, the additional time required for the sound wave to reach the far ear may be a fraction of the period of the wave, such as $\frac{1}{4}$ of the period. In this case, the corresponding interaural difference in signal *phase* would be $\frac{1}{4}$ of 360°, or 90°. Increments or decrements in the arrival time of near- and far-ear wave-forms would result in corresponding increases or decreases in the difference in interaural phase. If the position of the source of the sound is moved so that the arrival time difference is increased from a fraction of the period to exactly match the period of the signal, the near- and far-ear waves would again be in phase. The sound wave incident at the far ear would be precisely one cycle behind the corresponding wave at the near ear. In the special cases where the arrival time difference between the near- and far-ear waves happens to coincide with two-times, three-times, or other integral multiples of the period of the signal, the near- and far-ear waves will again be in register, and there will be no interaural differences in phase. Because it is unlikely that the source of a sound will be positioned such that the arrival time differences will equal the period (or multiples of the period) of the signal, interaural differences in arrival time usually result in interaural differences in phase, and these differences may be a prominent cue for localizing the azimuth of the source.

Interaural *level differences* are an additional cue for the perception of the azimuth of sound position. Level differences may occur when the origin of the sound is off to one side, and they are a consequence of the shape of the torso, head, pinna, and external ear canal and the properties of sound diffraction, reflection, and refraction with these structures. The magnitude of sound diffraction, reflection, and refraction is dependent upon the relative dimensions of the wavelength of the sound wave and the size of the reflective structures. In general, interaural level differences are most important for signals composed of wavelengths that are less than the diameter of the listener's head. Shorter wavelengths (e.g., higher frequency signals) produce more prominent interaural level differences, and the characteristics of these differences are highly dependent upon the specific geometry of the listener's head and pinna.

2.2 Geometrical Considerations

Rayleigh (1876, 1945) was the first to propose that the head be idealized as an acoustically opaque sphere with the ears diametrically opposed. With this idealization, the shape and dimensions of the pinna are ignored. Furthermore, the idealized model assumes that the head is immobile (unable to scan the sound field) and that a point sound source is positioned greater than one meter from the listener. Under these conditions, the wave front may be approximated by a plane. Interaural distance differences (Δd) will occur for all sound locations other than those which lie on the median plane. As depicted in Figure 3.1, for a sound source to the right of a listener at azimuth X, the additional distance that the sound must travel to reach the far ear (left ear) is given by the sum of the linear distance, r(sin X), and the curvilinear distance, r(X). That is, the difference (Δd) in the sound pathlength for the two ears is given by Equation 1,

$$\Delta d = r(X + \sin X) \qquad \text{Eq. 1}$$

where Δd is the distance difference in cm, r is the radius of the listener's head in cm, and angle X is measured in radians.

The pathlength difference to the two ears is acoustically realized by the interaural difference in time-of-arrival of corresponding points in the waveforms incident at the two ears. Time-of-arrival differences (Δt) are calculated by dividing the distance difference by the velocity of sound. Given a sound velocity in air of 343 m/s, then the relationship between Δt and azimuth is provided by equation 2,

$$\Delta t = r(X + \sin X)/3.43 \times 10^4 \qquad \text{Eq. 2}$$

where Δt is the temporal difference in μs, r is the radius of the observer's head in cm, and angle X is measured in radians.

Three factors merit emphasis. First, Δt approaches a maximum value as azimuth X approaches $\pi/2$ radians or $3\pi/2$ radians (90° or 270°). Assuming r = 8.75 cm, the usual value assigned for humans, then maximum Δt = 656 μs. That is, at $\pi/2$ radians (90°), the sound wave arrives at the far ear 656 μs after it arrives at the near ear. Second, for any given azimuth, Δt varies directly with r. As a result, listeners with large heads will experience a greater interaural time-of-arrival difference than will subjects with small heads. Consequently, if the neural processing resolution of time-of-arrival differences is approximately equal across mammalian species, then species with large heads will be able to perceive finer changes in azimuth using this cue than will species with small heads. Third, interaural time-of-arrival differences do not define a specific locus in three-dimensional space. That is, sources that differ in elevation may still have the same interaural pathlength difference. Furthermore, the hemifield behind a listener is a mirror image of that in front, and locations above, behind, or below a listener may have the same interaural time-of-arrival difference as those in the front hemifield.

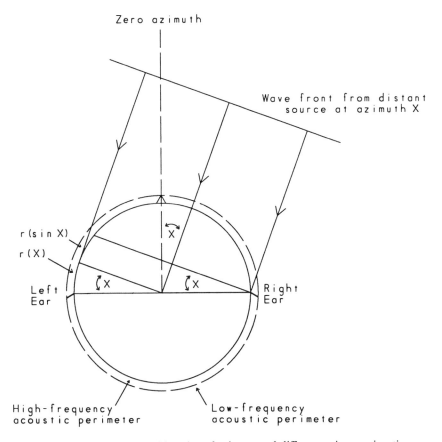

Zero azimuth

Wave front from distant
source at azimuth X

r (sin X)

r (X)

Left
Ear

Right
Ear

High-frequency
acoustic perimeter

Low-frequency
acoustic perimeter

FIGURE 3.1. Geometrical considerations for interaural differences in ongoing time or phase differences as a function of frequency. The signal is presented at azimuth X at a distant location from the listener. The interaural differences in signal phase obtained with low-frequency tones are produced by a larger effective acoustic radius of the head compared to that obtained by high-frequency signals. High and low frequency is scaled relative to head size. A high-frequency signal is one in which the wavelength of the sound is equal to or less than 2X the radius of the listener's head, while a low-frequency tone is one in which the wavelength is 8 times the radius of the head or greater. In the intermediate frequency zone (defined by the interval $2r < X < 8r$), the effective radius of the head gradually changes from the two boundary conditions illustrated here. See Kuhn (1977) for a detailed treatment of these phenomena.

2.3 Frequency and Effective Radius

Time-of-arrival cues are produced for both the leading edge of the wave front and for ongoing time or phase differences in the wave-forms. Interaural differences in time-of-arrival cues are frequency, head size, and azimuth dependent. Kuhn (1977, 1987) has shown that the effective acoustic radius of the head is larger than the skull perimeter when low-frequency stimuli are presented, but equal to the skull perimeter when high-frequency sounds are presented. In general, when the wavelength of the stimulus is less than or equal to the diameter of the skull (a high-frequency signal), the effective acoustic radius approximates that of the perimeter of the skull. When the wavelength of the stimulus is greater than or equal to four times the diameter of the skull (a low-frequency signal), the effective acoustic radius expands to a larger value with a magnitude that is probably governed by the magnitude of prognathism (the ventral protrusion of the jaw and nose), and by the size of the pinna. In humans the effective acoustic radius of the head for low-frequency signals is about 150% of that for high-frequency signals (Kuhn 1977), and it is probable that animals with pronounced prognathism of the nose and jaw and hypertrophied pinna would experience even greater expansion of this parameter for low-frequency signals. This phenomenon is not intuitively apparent, and it is attributed to a frequency dependence in the pattern of standing waves created around acoustic barriers (Kuhn 1977).

The functional acoustic radius for the leading edge of a sound wave, however, is equal to the skull perimeter, and it is not influenced by the relative frequency of the signal (Kuhn 1977). Thus, as depicted in Figure 3.1, low-frequency signals have an enhanced difference in interaural phase because the effective acoustic radius of the head is expanded for the fine structure of these signals.

The cues that are available for time domain processing are influenced by the rise and fall time of the signal and the complexity of the frequency spectrum and envelope (or amplitude modulation) of the wave-form. Signals which seem to begin and end imperceptibly have a slow rise and fall time (gradual onset and offset). These signals lack a crisp leading edge. Hence, time domain localization would likely be restricted to the comparison of interaural differences in the phase of the fine structure of the signal or to the comparison of interaural differences in the amplitude contour, or envelope, of the wave-form (Henning 1974; McFadden and Pasanen 1976). In the case of an unmodulated, slow onset and offset pure tone, time domain processing would necessarily be restricted to an analysis of the interaural phase differences of the fine structure of the tone. However, in spectrally and temporally complex signals, the envelope will be modulated, and the envelope of these modulations will be incident at the near and far ear at correspondingly different times-of-arrival. Human subjects are able to localize signals by processing interaural differences in signal envelopes (Henning 1974; McFadden and Pasanen 1976), and it is likely that these envelope cues influence sound localization in other mammals as well (Brown et al. 1980). Thus, time domain processing of localization cues may be directed at the analysis of interaural differences of the cycle-by-cycle fine structure of the signal or at the analysis of interaural differences in the time-of-arrival of the more global modulations of the envelope of complex signals (Middle-brooks and Green 1990).

2.4 Azimuth Ambiguity

Interaural differences in signal phase may provide ambiguous information regarding the position of the source. By way of example, assume that the radius of the head is 8.75 cm and the maximum time difference is 656 μs for the fine structure of the signals in question. In this example, as a simplification, ignore the fact the effective acoustic radius may change for signals of different frequency. This interaural time difference ($\Delta t = 656 \mu$s) would result in interaural phase differences of 90°, 180°, and 360° for pure tones of 380 Hz, 760 Hz, and 1520 Hz, respectively. This example illustrates two points. First, the relationship between interaural phase difference and spatial location is frequency dependent. A phase difference of 30° indicates one position for a tone of one frequency, but a different position for a tone of another frequency. Second, more than one location may produce the same difference in interaural phase when the period of the wave-form is equal to or less than twice the maximum interaural difference in time-of-arrival. In this simplified example for human listeners, such locational ambiguities will occur for frequencies with periods less than or equal to 1312 μs. Here, a 760-Hz stimulus will produce a 180° difference in interaural phase when the stimulus is presented either at azimuth $\pi/2$ radians or $3\pi/2$ radians (90° to the right or left). Hence, interaural phase information alone will not discriminate between these two locations. Similarly, for all frequencies greater than 760 Hz, the interaural difference in signal phase produced for a source at any given azimuth will be perfectly matched by at least one other azimuth. The possibility of ambiguity in azimuth for interaural phase differences of midrange and high-frequency signals suggests that phase information should be utilized in sound localization only for low-frequency signals. Furthermore, the smaller the head size, the higher the frequency limit for unambiguous localization via interaural phase. A small rodent with a maximum Δt of only 100 μs will not experience ambiguous azimuths for phase differences of signals below 5000 Hz in frequency.

The perception of interaural differences in the phase of the fine structure is restricted to relatively low-frequency signals. Both physiological and behavioral observations indicate that the mammalian auditory system is unable to resolve interaural differences in signal phase for frequencies above some critical value. The critical value may differ for various species, and it is usually observed in the region of 1 kHz to 5 kHz (Klumpp and Eady 1956; Kiang et al. 1965; Rose et al. 1967; Anderson 1973; Brown et al. 1978a; Johnson 1980). It is likely that the critical value observed in different studies reflects the operation of two factors: the upper limit for phase locking and the ambiguous azimuth factor. Different researchers use different methodologies and criteria for estimating the upper limit of sensitivity for interaural phase differences (Kiang et al. 1965; Rose et al. 1967; Anderson 1973; Brown et al. 1978a; Johnson 1980); nevertheless, the data generally support the idea that mammals with small heads are sensitive to interaural phase differences for higher frequency signals than are mammals with larger heads. Because the ambiguous azimuth factor may lead to costly localization errors, it is possible that mammals with large heads have been selected to lower the upper frequency limit of interaural phase sensitivity below the upper limit that fibers in the auditory nerve are capable of phase locking.

Though interaural differences in time-of-arrival may be described rather accurately by the geometrical model presented above, the same is not true for interaural differences in signal level. In the case of interaural level differences, subtle variations in the shape of the skull and pinna have a pronounced

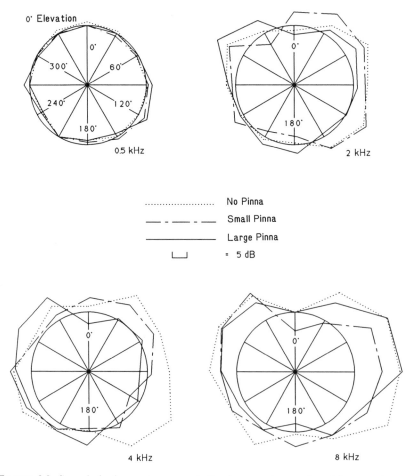

FIGURE 3.2. Sound shadows at representative frequencies produced by rotating an artificial head fitted with large, small, or no pinnae. Points further from the center of the circle indicate that the signal level was more intense at the ear on that side of midline. (From Harris 1972, reprinted with permission.)

impact on the magnitude of the observed differences in interaural level. Using a Shilling artificial human head, Harris (1972) conducted measurements of interaural level differences with either no pinna, a large pinna, or a small pinna, chosen to sample human pinna variations. These studies were conducted with a microphone diaphragm placed in the position of the tympanic membrane, and differences in the sound pressure level incident at each eardrum were measured as a function of the azimuth of the sound source, the frequency of the signal, and the size of the pinna (large pinna, small pinna, or no pinna). Harris' measurements, presented in Figure 3.2, show that at low frequencies (e.g., 500 Hz) interaural level differences were very small, while at high frequencies (e.g., 8 kHz) level differences were prominent. The results also indicate that this is an area where mathematical models do not substitute for empirical measurements. For example, it is surprising that, at some azimuth and frequency combinations, interaural level differences were greater for the no pinna condition than they were for either the large or small pinna conditions.

Harrison and Downey (1970) used probe microphones placed by the tympanic membrane to measure interaural level differences in humans, rats, bats, and squirrel monkeys. Their data showed that interaural level differences tended to increase with frequency, and they encountered very large interaural level differences with nonhuman subjects. Figure 3.3 displays interaural level differences for an individual squirrel monkey (*Saimiri sciureus*). In general, as the signal frequency was increased, interaural level differences also increased and, at certain azimuth and frequency combinations, could exceed 20 dB.

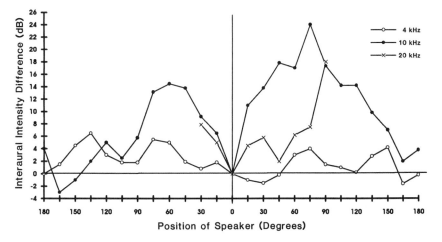

FIGURE 3.3. Interaural level differences measured in squirrel monkeys (*Saimiri sciureus*) as a function of the position of speaker azimuth (0° to 180° on either side of zero azimuth) at three tone frequencies. (From Harrison and Downey 1970, reprinted with permission.)

However, because the magnitude of interaural level differences was influenced by small variations in the morphology of the head and pinna, interaural level differences do not vary monotonically with changes in azimuth. It is possible that with tonal, or narrow bandwidth, signals two or more azimuths may give rise to the same overall interaural level difference, and sound position may then be ambiguous. Broad bandwidth, high-frequency signals may be accurately localized via the interaural level difference mechanism, however. Brown et al. (1978a) have argued that at each azimuth the left and right ears will have a spectral transfer function, and the difference between the near- and far-ear functions will give rise to a binaural difference spectrum (Fig. 3.4).

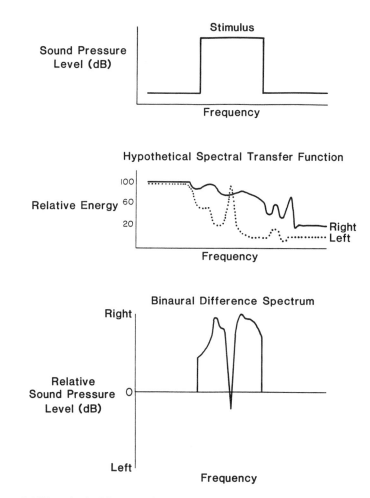

FIGURE 3.4 Hypothetical interaural sound pressure level differences as a function of the azimuth of the source and signal frequency. Negative level differences in the difference spectrum are generated when the signal level incident at the far ear exceeds that recorded at the near ear. (From Brown et al. 1978a, reprinted with permission.)

The properties of the binaural difference spectrum may be unique to each azimuth, and if the stimulus is broadband, then accurate sound localization would be possible.

In many groups of mammals, the localization of sound azimuth may be dependent upon the perception of interaural differences in time-of-arrival and in signal level. At signal frequencies above which interaural time-of-arrival differences become ambiguous, the interaural level difference may become a viable cue. Thus, sound localization may be governed by a *duplex perceptual system*. It is possible that some mammals may be more dependent upon one mechanism, while other mammals are more dependent upon the other. Furthermore, it is likely that head size differences and pinna size morphology may amplify the significance of one mechanism relative to that for the other.

2.5 The Vertical Coordinate of Sound Location

The perception of source height, or elevation, may be particularly significant for the arboreal primates, marine mammals, and other nonterrestrial organisms. At midcanopy levels in the rain forest and in marine habitats, biologically significant sounds may arise from positions above and below the listener, as well as from the right and left. If the listeners were unable to move their pinnae and if their right and left ears were acoustically symmetrical, then vertical localization would not involve binaural processing, unless, of course, the listeners strategically cocked their heads (Menzel 1980). The left and right elevation-dependent sound transformation functions are not bilaterally symmetrical in cats (Musicant, Chan, and Hind 1990), humans (Shaw 1974; Middlebrooks, Makous, and Green 1989), and barn owls, *Tyto alba* (Payne 1962; Norberg 1977; Knudsen, Blasdel, and Konishi 1979). Binaural asymmetries help distinguish variations in elevation from variations in the frequency composition of the signal.

In human listeners (and probably in most terrestrial mammals), the perception of vertical position is dependent upon the fact that the transformation function of the external ear is elevation dependent. Using high-frequency, broad bandwidth signals, the apparent spectral content of the sound changes with changes in the elevation of the stimulus (Butler 1969; Gardner 1973; Hebrank and Wright 1974; Kuhn 1979). For changes in the "apparent" spectral content to serve as a cue for elevation, the frequency composition of the signal must be "stable" and the listener must be "familiar" with the characteristics of the signal, or the elevation-dependent transformation functions of the two ears must be sufficiently distinctive that the elevation of a novel sound may be discerned through interaural comparisons. Because the asymmetries and convolutions of the pinna and external ear canal must be relatively large compared to the wavelength of the signal for the expression of elevation-dependent differences in the external ear transformation function, this cue would require relatively high-frequency, broad bandwidth signals and high-frequency hearing (Shaw 1974; Kuhn 1979; Wightman and Kistler 1989a,b). However, lower frequency signals may reflect off the ground and the organism's torso in an elevation-dependent manner (Kuhn 1979; Brown et al. 1982), and it is possible that some degree of vertical localization is possible with low-frequency signals. Nearly all mammals have excellent high-frequency hearing (for a review see Long, Chapter 2), and this general trait in many cases may be at least as critical for vertical localization as it is for horizontal localization.

While humans have short, essentially fixed pinnae, most terrestrial mammals have extended, mobile pinnae, and asymmetries in pinna shape or orientation (Searle et al. 1975) may enhance vertical localization just as ear canal asymmetries enhance the perception of acoustic elevation in the barn owl *Tyto alba* (Payne 1962; Norberg 1977; Knudsen, Blasdel, and Konishi 1979). Though marine mammals have either no pinna or a small pinna, accurate vertical localization may still be possible (Renaud and Popper 1975). Much work remains to be conducted regarding marine mammal localization; it is unknown how sound is propagated around the head, torso, and ear canal of various marine mammals, and it is unknown if left-right asymmetries exist in the transformation functions for sounds presented at different elevations.

2.6 The Distance Coordinate of Sound Location

The perception of acoustic proximity (distance or depth) is very poorly understood, yet its analogue has been well studied in the visual system. In visual perception, both binocular and monocular cues may provide information regarding the relative proximity of visual targets. The chief binocular cue is binocular disparity: a near object is seen from two slightly different angles by the two eyes. When the observer views a near object, more distant objects in the background will necessarily fall on different areas of the left and right retinas. A second binocular cue is convergence, the inward turn of the eyes required to maintain stereoscopic vision. This cue becomes more pronounced as the visual target is positioned progressively closer to the subject (Kaufman 1979). Because there is very little change in either of these binocular cues for two targets positioned at 10 m or 20 m, for example, it is likely that relative distance judgments for distal targets are more dependent upon monocular cues than on binocular ones.

The monocular cues for distance perception (for a review see Kaufman 1979) include size constancy (the retinal image size varies with changes in object distance), interposition (near objects are in front of, or partially obscure, more distant objects), linear perspective (parallel lines appear to converge at the horizon), textural perspective (the density of items per unit of retinal area increases with distance; that is, more trees would be visible in a forested hillside if the hillside was viewed at progressively greater distances), aerial perspective (distant objects appear to lose their color saturation and appear to be tinged with blue), relative brightness (objects at greater distances from a light source, such as a street light, have less luminance than do objects positioned closer to the source), and relative motion parallax (the apparent location of distant objects is shifted less by a change in the position of the viewer than by the perceived locations of closer objects).

In light of the richness of the appreciation of the cues underlying visual depth and distance perception, it is surprising that so little is known about either the putative perceptual cues underlying the perception of acoustic distance or the abilities of various species to perceive differences in the proximity of acoustic events. Nevertheless, there is good reason to believe that the perception of acoustic proximity has undergone intense selection in a broad array of organisms. Payne (1962) showed that, in a totally darkened room, barn owls (*Tyto alba*) were able to accurately fly from an elevated perch to broadcasts of recordings of the rustling noise produced by the movements of a mouse. Because the barn owl flies headfirst, yet captures prey feetfirst, it must be able to accurately estimate the acoustic azimuth, elevation, and distance to be able to position its body for the strike. If the owl were unable to perceive acoustic proximity, it would risk breaking its descent either too soon or too late. More recently, playback experiments have shown that great tits, *Parus major* (McGregor and Krebs 1984; McGregor, Krebs, and Ratcliffe 1983), howler monkeys, *Aloutta palliata* (Whitehead 1987), and grey-cheeked mangabeys, *Cercocebus albigena* (Waser 1977), use acoustic cues to gauge distance and judge the possibility of incursions into one's territory by a rival individual or group. It is likely that the perception of acoustic distance is important to many species.

Though binocular vision is important for distance and depth perception, there is little available evidence to suggest that binaural hearing is either important, or unimportant, for the perception of acoustic proximity. It is likely that many of the monocular and binocular cues for distance perception have a rough analogue in the acoustic domain.

The prime candidates for monaural distance perception include:

1. Amplitude, sound level, or auditory image constancy (the amplitude of the signal varies with distance usually in accordance with the inverse square law (Gamble 1909; Coleman 1963). Hence, the raucous calls of the hornbill, *Bycanistes subcylindricus*, grow softer as the bird flies to a more distant part of the forest.
2. Frequency spectrum at near distances (von Békésy 1938). At distances of less than four feet, the low-frequency components of complex signals are relatively more prominent than are the midfrequency and high-frequency components, and as the source of the signal is moved progressively closer to the listener, the low-frequency components become even more prominent.
3. Frequency spectrum at far distances (Hornbostel 1923; Coleman 1963). The molecular absorption coefficient for sound in air is humidity, temperature, and frequency dependent. At a temperature of 20°C and a relative

humidity of 50%, the absorption coefficient of a 10-kHz tone is about 20-fold greater than that for a 1-kHz tone (Nyborg and Mintzer 1955). Hence, high frequencies are attenuated more rapidly than are low frequencies, and at successively greater transmission distances, the frequency spectrum of complex signals shows a progressive loss of the high-frequency components (Waser and Brown 1986). This cue resembles the aerial perspective cue in the visual domain. That is, just as more distant views are characterized by the loss of longer wavelength hues, more distant sounds are characterized by the loss of high-frequency components.

4. Reverberation. The temporal patterning of signals becomes "smeared" as the delayed reflected waves overlay the direct wave (Mershon and Bowers 1979). Hence, the ratio of direct-to-reflected waves can provide distance information. This phenomenon is more likely to provide usable information in forested habitats than it is in open habitats.

5. Temporal distortion. Changes in wind velocity, wind direction, or convection current flow result in changes in the duration and pitch of signals transmitted through an unstationary medium. Signals broadcast from greater distances are probably more susceptible to disturbance by this phenomenon, but this has not been studied in detail (Brown and Gomez 1992).

6. Movement parallax. The relative location of distant sources is shifted less by a change in location of a listener than by the perceived locations of closer sources. This cue is a direct analogue to relative motion parallax in the visual domain. It is probable that this cue requires rather large displacements in space for it to play a role in distance judgments for head cocking, and other rotational movements of the head and neck may be insufficient to aid distance judgments in some situations (Simpson and Stanton 1973).

Binaural cues for the perception of acoustic distance include:

1. Binaural intensity ratio. When the source of a signal is at a position other than $0°$ azimuth, the signal may be greater in amplitude at the near ear relative to the amplitude of the signal at the far ear. This difference in sound amplitude, the binaural intensity ratio, varies as a function of head size, azimuth, signal frequency, and transmission distance (Hartley and Fry 1921; Firestone 1930).

2. Binaural differences in the signal phase. In addition to the binaural intensity ratio, empirical measurements have shown that binaural differences in the signal phase vary as a function of transmission distance as well as of head size, azimuth, and signal frequency (Hartley and Fry 1921). Thus, it is possible that binaural differences in the signal phase may help cue transmission distance.

3. Acoustic field width. At the front row of the auditorium, the orchestra may occupy a whole hemifield, while at the rear of an auditorium, the orchestra occupies a more restricted portion of the acoustic field. Hence,

the perceived distance to an acoustic source that is not a point source varies inversely with the perceived width of the acoustic field (Brown and Brown in preparation). Though this putative cue is likely binaural in origin, it resembles the monocular cue of textural perspective in the visual domain.
4. Scattered sound direction and field width. In forested habitats, sound becomes scattered by tree trunks. The greater the transmission distance, the greater the magnitude of the field width of the scattered sound, and the perceived width of the field of this scatter may influence distance judgments (Brown and Brown in preparation). There is very little data to indicate the relative potency of the various putative monaural and binaural cues for judgments of distance, and much research remains to be done in this area.

The utility of these cues for the perception of acoustic proximity depends on how reliably they change with changes in distance. The initial cue listed above, auditory image constancy, is simply a change in signal amplitude, while all the other cues enumerated here are associated with a change in sound quality, sound distortion, or sound characteristics at each ear. The only cue which has received full examination is auditory image constancy (e.g., amplitude constancy) (Gamble 1909; Ashmead, LeRoy and Odom 1990); however, studies of sound transmission in natural habitats have shown that the amplitude may fluctuate 20 dB or more in short intervals of time (Wiley and Richards 1978; Waser and Brown 1986). Fluctuations of this magnitude may lead to errors in judgment of three to four doublings of acoustic distance (a sound presented at 25 m under unfavorable conditions may be received at a lower amplitude than the same sound at 100 m presented under favorable conditions). Hence, sound amplitude per se is generally regarded of as a poor index of transmission distance (Brown and Brown in preparation).

In probably all habitats, sounds are degraded by the natural environment, and these more complicated habitat-induced changes in sound quality may more reliably cue acoustic proximity. Brown and Waser (1988) have shown that different primate calls are degraded differently by the acoustics of natural habitats. Changes in sound quality have been measured with respect to the frequency composition of the call and the temporal patterning of the signal (Brown and Waser 1988; Brown and Gomez 1992).

Figure 3.5 shows sound spectrograms of the blue monkey (*Cercopithecus mitis*) grunt utterance at the source (panel A) and three rerecordings of the same call after having been broadcast 100 m in the savanna (panels B to D). While the signal displayed in panel B retains the overall structure of the source (panel A), the signal shown in panel C is missing the low-frequency portion of the call (the band of energy at about 500 Hz), and the signal displayed in panel D is missing the two higher frequency components of the call (the bands of energy at about 1500 Hz, and 3000 Hz). These recordings

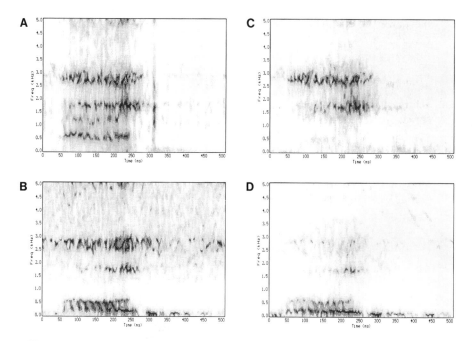

FIGURE 3.5. Sound spectrograms of a grunt call given by a blue monkey (*Cercopithecus mitis*). Panel A illustrates the call at the source. The signal is composed of three energy bands centered at approximately 500 Hz, 1500 Hz, and 2500 Hz. Panels B to D display spectrograms of the call recorded at a transmission distance of 100 m in the savanna habitat. Panel B: the recording was noisy but all elements of the call were present. Panel C: the 500-Hz frequency band was absent. Panel D: the 1500-Hz and 2500-Hz frequency bands were strongly attenuated. (From Brown and Gomez 1992, reprinted with permission.)

were conducted in succession under rather stable climatic conditions within a two-hour interval at the same site, and a review of the factors in natural habitats that lead to different patterns of distortion are beyond the scope of this paper (for a review see Brown and Gomez 1992). Nevertheless, these recordings dramatize the fact that the structure of signals may be altered by the acoustics of the habitat. Environmentally induced degradation of acoustic signals occurs in probably all natural settings, and some types of distortion may be useful for estimating the distance of the source.

It is possible to adopt signal processing techniques to quantitatively measure the magnitude of habitat-induced distortion of vocalizations (Brown and Waser 1988; Brown and Gomez 1992). The data show that some vocalizations (e.g., the blue monkey's boom) are relatively unchanged by the acoustics of the habitat, while other calls (e.g., the blue monkey's chirp or pyow) are more susceptible to degradation. The overall pattern of these degradation scores indicate that different utterances are degraded in different ways by environmental acoustics. This finding suggests that some vocalizations may be good for revealing acoustic proximity, while other utterances may obscure the relative proximity of the vocalizer. Presumably, the presence or absence of "distance information" in various calls is relevant to the social function of different vocalizations. Many forest monkeys emit calls that appear to mark the position of the vocalizer. These calls may be involved in regulating the spacing, distribution, and movements of individuals out of visual contact.

3. Sound Localization Methodology

Many animals will orient towards, and approach, the site of origin of some sounds. The accuracy of approach has been used to study sound localization in the grey-cheeked mangabey monkey *Cercocebus albigena* (Waser 1977), tree frogs *Hyla cinera* and *H. gratiosa* (Feng, Gerhardt, and Capranica 1976), cats (Casseday and Neff 1973), and many other species. In some instances, food or some other reward has been used to maintain this behavior. In such approach procedures, the accuracy of localization is dependent on the ability of the auditory system to process a change in sensation associated with a change in the position of the source and on the ability of the motor systems of the animal to accurately guide the subject towards the perceived location of the acoustic target. Species differences in the acuity of localization, measured by the approach procedure, may be due to differences in the precision of the perceptual system or, alternatively, these apparent acuity differences may be due to variations in the accuracy of motor systems.

Orientation paradigms have also been developed to measure the acuity of localization. With these methods, a head turn or a body turn is used to indicate the perception of sound direction (Knudsen and Konishi 1978; Knudsen, Blasdel, and Konishi 1979; Whittington, Hepp-Reymond, and Flood 1981; Brown 1982a; Perrot, Ambarsoom, and Tucker 1987; Makous and Middlebrooks 1990). Both orientation and approach procedures are categorized as egocentric methods (Brown and May 1990). Here, localization is made not with reference to an external acoustic marker, but rather with reference to the subject's physical orientation in space. With egocentric procedures, differences in sound localization acuity may be due to limitations in the accuracy of the perceptual system or to limitations of the motor system.

Behavioral tasks in which listeners have been trained to operate response levers to indicate the detection of a change in sound location have been used with both human (Mills 1958) and animal subjects (Brown et al. 1978a). These ear-centered, or otocentric, procedures are designed to directly assess the acuity of the perceptual system, and they do not require the participation of the spatial/motor system (Brown and May 1990). Hence, with these procedures, listeners report the detection of a change in sound location, but they do not indicate where the sound originated relative to their own orientation.

Given the procedural variations possible between these different method-

FIGURE 3.6. Sound-initiated approach in grey-cheeked mangabeys (*Cercocebus albigena*) evoked by the playback of a whoopgobble vocalization. P1 is the location of the broadcast, and angle θ depicts the average error of approach to the broadcast site. The filled circles denote explicit sightings of the path of approach. The playback was conduced in the Kibale forest in Uganda with native populations of mangabeys. (From Waser 1977, reprinted with permission.)

ologies, it is important to note that independent measurements of sound localization acuity in normal animals appear to be remarkably consistent and robust. There is a high degree of agreement in the results using both egocentric and otocentric methods within (Heffner and Heffner 1988d) and between laboratories (Brown and May 1990). Using the approach procedure under field conditions in the natural habitat, Waser (1977) showed that one of the Cercopithecoidae monkeys, the grey-cheeked mangabey (*Cercocebus albigena*), was able to localize the whoopgobble, a complex long-distance vocalization (Brown 1989), with an average error of only 6° (Fig. 3.6). Under laboratory conditions, using otocentric methods with two other species of the Cercopithecoidae monkeys (*Macaca nemestrina* and *M. mulatta*), the localization of complex vocal signals ranges from 3° to 15° depending upon the specific acoustical characteristics of the utterance (Brown et al. 1978b, 1979). The mean localization error of macaque monkey broad bandwidth or frequency-modulated calls, those which are most comparable to the mangabey's gobble, is about 3°. It is remarkable that a field phonotaxis study conducted in the monkey's native habitat in Uganda and a laboratory investigation yield results that are so similar.

When comparable stimuli are used, the congruence in the data produced by different laboratories employing different methods is even more striking. Figure 3.7 shows interaural time difference thresholds measured using earphones (Houben and Gourevitch 1979) and those calculated from free-field measurements of the acuity of localization (Brown et al. 1978a) as a function of tone frequency. Data for human subjects (Klumpp and Eady 1956) are compared with macaque monkey data. These data show that the physical characteristics of the signal have a strong impact on the accuracy of sound localization. This is true for both simple synthetic signals such as tones and complex natural signals such as vocalizations. Furthermore, the data show

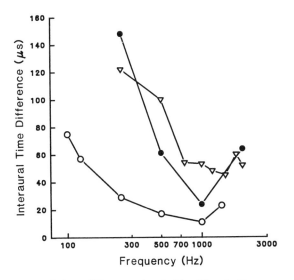

FIGURE 3.7. Interaural time difference thresholds. Thresholds, measured in micro-seconds, are displayed as a function of signal frequency. Macaque monkey (*Macaca mulatta* and *M. nemestrina*) thresholds (black circles) are transposed from free-field localization measurements (Brown et al. 1978a). Interaural time difference thresholds were measured with earphones in monkeys (open triangles; Houben and Gourevitch 1979) and humans (open circles; Klumpp and Eady 1956). (From Brown et al. 1978a, reprinted with permission.)

that measurements of a species' acuity for sound localization are robust and relatively independent of method. These observations indicate that it is possi-ble to measure with high precision the acuity of sound localization which is representative of the abilities of the species, but that the data derived are dependent upon the physical characteristics of the test signal.

4. The Acuity of Sound Localization

4.1 The Perception of Acoustic Azimuth

The just detectable change in the position of the sound source, the minimum audible angle, has generally been regarded as the most quantitative index of the acuity of localization. Figure 3.8 presents individual psychometric sound localization functions for three macaque monkeys. The test signal was a macaque coo vocalization. The psychometric functions were derived from monkeys who had been trained to hold a contact-sensitive key when sounds were pulsed repetitively from a source at 0° azimuth, directly in front of the monkey, and release contact with the key when the sound was pulsed from a source at any other azimuth. The monkey's rate of guessing (its catch-trial

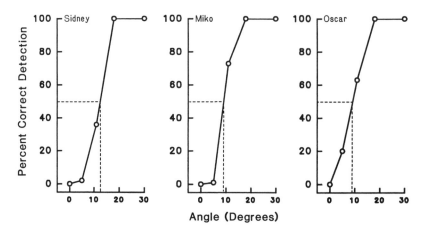

FIGURE 3.8. Psychometric functions for the localization of a macaque coo call. Functions are shown for three individual monkeys (Sidney, Miko, and Oscar). The monkey's rate of guessing (catch-trial rate) is displayed over the zero-degree point, and the monkey's percent of correct detection for the trials presented at the four comparison locations increased with angle. The calculation of the minimum audible angle is shown by the dashed line. (From Brown et al. 1979, reprinted with Permission.)

rate) was very low, less than 8% (this rate is displayed over the 0° point in Fig. 3.8). The monkey's ability to detect a change in the azimuth of the sound source increased with the magnitude of change in source location, reaching about 100% correct by 30°. These psychometric functions conform to the classic ogive shape (Cain and Marks 1971), and the 50% correct detection point (the minimum audible angle) is measured in degrees and calculated from the psychometric functions.

Investigators have tended to measure the acuity of directional hearing with biologically significant stimuli such as vocalizations (Feng, Gerhardt, and Capranica 1976; Waser 1977; Brown et al. 1978a, 1979; Rheinlaender et al. 1979) or, more commonly, with synthetic signals that are either simple, such as pure tones (Casseday and Neff 1973; Terhune 1974; Brown et al. 1978a; Heffner and Heffner 1982), or complex, such as clicks or noise bursts (Ravizza and Masterton 1972; Brown et al. 1980; Heffner and Heffner 1982, 1983, 1987, 1988a,b). Biologically significant signals have tended to be used with phonotaxic procedures or in studies in which the relative acuity of localization of various natural signals is the topic of interest, while synthetic signals have tended to be used in studies which have focused on assessing the limits of the perceptual system.

4.2 Biologically Interesting Signals

The different physical characteristics of various classes of complex natural stimuli, such as vocalizations, may influence the acuity of localization. In

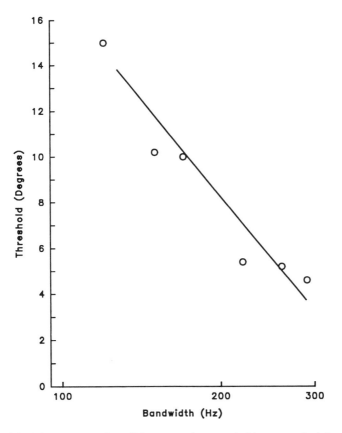

FIGURE 3.9. Macaque monkey (*Macaca mulatta* and *M. nemestrina*) localization thresholds for six coo vocalizations displayed as a function of the effective bandwidth (frequency-modulated bandwidth) of the dominant band of the call. The correlation between threshold and call bandwidth was −0.98. (From Brown et al. copyright 1978b, by the AAAS. reprinted with permission.)

macaque monkeys, the effective bandwidth (or magnitude of frequency modulation) of the dominant frequency band of the call has a strong effect on sound localization (Fig. 3.9). The greater the effective bandwidth, the greater the accuracy of localization. Minimum audible angles for macaque coo calls span approximately a fivefold range, from about 3° to 15°. Macaque monkeys also produce a wide variety of noisy barks, grunts, and growls, and these harsh sounding, atonal, broad bandwidth calls are all accurately localized as well (Brown et al. 1979). Complex natural signals that exhibit a broad effective bandwidth (produced either by frequency modulating a relatively tonal sound or by generating an atonal, broad bandwidth sound) are probably localized at the limits of resolution of the organism's perceptual system (Brown 1982b). The mate-attracting calls, rallying calls, and position-marking calls given by a wide variety of mammals typically exhibit a broad effective bandwidth, and this attribute likely promotes sound localization at the listener's limit of resolution.

4.3 Pure Tones

Comparative data for the localization of pure tones as a function of frequency are shown in Figure 3.10. While the human data suggest that the stimulus frequency has a relatively modest effect on the localization of tones (Mills 1958), it tends to have a pronounced effect on the accuracy of localization by nonhuman mammals. At the best frequency, macaque monkeys, *Macaca mulatta* and *M. nemestrina* (Brown et al. 1978a), harbor seals, *Phoca vitulina* (Terhune 1974), and elephants, *Elephas maximus* (Heffner and Heffner 1982), exhibit a resolution of about 4°, while human listeners are able to resolve about 1° (Mills 1958). At the worst frequency, human subjects are still able to resolve angles of about 3°, while most of the other mammals tested may require angles of 20° or more. Thus, human subjects tend to be more accurate at localizing the source of sounds across the frequency spectrum than are most other mammals. Testing with pure tones has almost exclusively been conducted with signals that are gated on and off slowly and are not modulated in amplitude so that the envelopes of the wave-forms do not

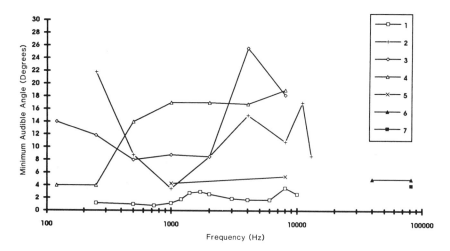

FIGURE 3.10. Sound localization acuity (minimum audible angles) of tones as a function of signal frequency for representative mammals. 1, human *Homo sapiens* (Mills 1958); 2, rhesus and pig-tailed macaque monkeys *Macaca mulatta* and *M. nemestrina* (Brown et al. 1978a); 3, cat *Felis catus* (Casseday and Neff 1973); 4, elephant *Elephas maximus* (Heffner and Heffner 1982); 5, harbor seal *Phoca vitulina* (Terhune 1974). 6 little brown bat, *Myotis Oxygnathos* (Ayrapet'yants and Konstantinov, 1974); 7 greater horseshoe bat, *Rhinolophos* ferrumequinum (Ayrapet'yants and Konstantinov 1974) (From Fay 1988, adapted with permission.)

provide information that may influence localization. Under these conditions, human listeners localize low-frequency tones with a mechanism sensitive to interaural time differences, while the localization of high-frequency tones is governed by a mechanism sensitive to interaural level differences (Mills 1960). The same frequency effects have been shown to hold for monkeys, *M. mulatta* and *M. nemestrina* (Brown et al. 1978a; Houben and Gourevitch 1979), and are presumed to apply for most other mammals as well.

The literature on the comparative localization of tones in the horizontal plane suggests that both mechanisms for localization by human subjects are equally accurate (Mills 1958, 1960), while in most other mammals one mechanism may be less accurate, and perhaps less significant, than the other. In this context, an extensive physiological and anatomical literature (Heffner and Masterton 1990) has shown that high-frequency localization primarily involves brainstem structures in the medial nucleus of the trapezoid body (MTB) and lateral superior olive (LSO), while low-frequency localization primarily involves structures in the medial superior olive (MSO). The relative development of these nuclei varies across mammals; in some species, such as humans and elephants (*Elephas maximus*), the MTB-LSO system is undeveloped or nearly absent, while in other species, such as the opossum (*Didelphis virginia*) and hedgehog (*Paraechinus hypomelas*), the MSO system is undeveloped or nearly absent. In general, as the physical size of the mammal increases, the greater the development of the MSO system and the concomitant reduction in the MTB-LSO system (Heffner and Masterton 1990). Thus, variations in the development of the auditory structures in the ascending pathway may underlie species differences in their ability to fully utilize interaural time-of-arrival difference cues or interaural level difference cues. These variations may account for the observed species differences in the pure tone localization data (Fig. 3.10). However, while human subjects localize high-frequency tones well, their MTB-LSO system is only marginally developed. Hence, although it appears that much is understood in the anatomical and physiological mechanisms subserving sound localization, significant puzzles still remain.

4.4 Complex Synthetic Stimuli

Comparative data for the localization of complex synthetic stimuli (e.g., clicks and noise bursts) are displayed in Figure 3.11. Here the minimal audible angle is plotted in reference to head size. As noted in Section 2.2, all other things being equal, both interaural time differences and interaural level differences should increase with head size. Thus, large mammals should exhibit greater sound localization acuity simply because the physical magnitude of these interaural cues increases with head size. This trend is generally observed (Fig. 3.11). However, the correlation between threshold and head size is only -0.59. Hence, some mammals are either significantly less sensitive, or more sensitive, to sound direction than would be expected by the size of their

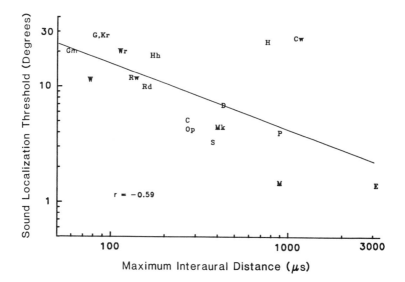

FIGURE 3.11. Interaural distance and sound localization acuity in representative mammals. Acuity is displayed for a broad bandwidth noise or click stimulus. Gm, grasshopper mouse *Onychomys leucogaster* (Heffner and Heffner 1988a); W, least weasel *Mustela nivalis* (Heffner and Heffner 1987); G, gerbil *Meriones unguiculatus* (Heffner and Heffner 1988c); Kr, kangaroo rat *Dipodomys merriami* (Heffner and Masterton 1980); Rw, wild Norway rat *Rattus norvegicus* (H. E. Heffner and R. S. Heffner 1985); Rd, domestic Norway rat and Wistar albino rat *R. norvegicus* (Kelly 1980); Wr, wood rat *Neotoma Floridiana* (Heffner and Heffner 1988a); Hh, hedgehog *Paraechinus hypomelas* (Chambers 1971); C, cat *Felis catus* (Heffner and Heffner 1988d); Op, oppossum *Didelphis virginiana* (Ravizza and Masterton 1972); S, harbor seal *Phoca vitulina* (Terhune 1974); Mk, rhesus and pig-tailed macaque monkeys *Macaca mulatta* and *M. nemestrina* (Brown et al. 1980); D, dog *Canis canis* (H. E. Heffner unpublished); H, horse *Equus caballus* (Heffner and Heffner 1984); M, human *Homo sapiens* (Heffner and Heffner 1984); P, domestic pig *Sus scrofa* (Heffner and Heffner 1989); Cw, cattle *Bos taurus* (Heffner and Heffner 1983); E, elephant *Elephas maximus* (Heffner and Heffner 1982). The regression between localization threshold and head size was −0.59, and the linear regression line is displayed by the solid diagonal line. (From Heffner and Masterton copyright © 1990, reprinted with permission of John Wiley & Sons, Inc.

heads. Species located below the diagonal regression line shown in Figure 3.11 have better localizational acuity than would be expected by their head size, while those positioned above the regression line have less acute directional hearing than would be expected. Thus, regardless of the magnitude of the physical cues available for localization, some species are particularly good localizers, while others are not.

How can these differences in the relative acuity for directional hearing be explained? In a classic paper, Harrison and Irving (1966) argued that sound localization abilities for many species may be important for acoustically redirecting the site of gaze. The horizontal width of the field of high-acuity vision is much narrower in animals with high-acuity binocular visual systems than in animals with nonoverlapping hemispheric visual systems. Heffner and Heffner (1988c) have reported a high correlation (r = 0.96) between the width of the zone of high-acuity vision and the relative sound localization acuity of representative species. If this correlation remains strong as a larger variety of species are examined, it would strengthen the idea that one key function of directional hearing systems is to acoustically guide the orientation of the visual system.

4.5 The Perception of Acoustic Elevation

The literature is much more limited concerning the comparative perception of acoustic elevation. In arboreal living species, or in marine mammals, the determination of acoustic elevation may be as significant as the determination of azimuth. Vertical and horizontal minimum audible angles for primate vocalizations are shown for macaque monkeys (*M. mulatta* and *M. nemestrina*) in Figure 3.12. The test vocalizations were a macaque coo call and a macaque grunt call. The grunt, which is broader in bandwidth, was localized more accurately than the coo. The median vertical localization thresholds were approximately 9° and 20°, respectively. For these same signals, the acuity of vertical localization was approximately two to three times less accu-

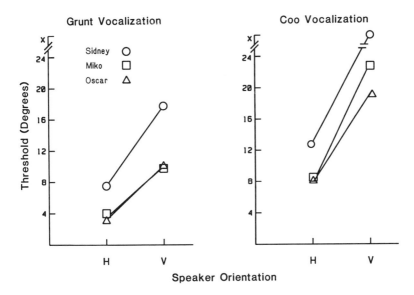

FIGURE 3.12. Horizontal (H) and vertical (V) minimum audible angles for a macaque grunt vocalization and a macaque coo vocalization for three macaque monkeys (Sidney, Miko, and Oscar). An X indicates that the performance level of the subject never exceeded chance. (From Brown et al. 1982, reprinted with permission.)

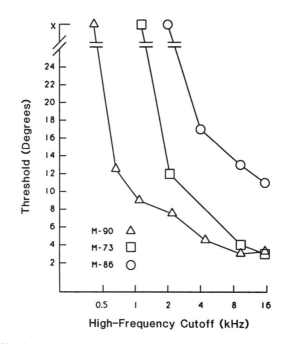

FIGURE 3.13. Vertical minimum audible angles for bandlimited noise as a function of the high-frequency cutoff of the noise band for three macaque monkeys (M-90, M-73, and M-86). The low-frequency limit of the noise was 125 Hz. An X indicates that the performance level of the subject did not exceed chance. (From Brown et al. 1982, reprinted with permission.)

rate than was localization in the horizontal plane. High-frequency hearing and high-frequency broadband stimuli are important for accurate vertical localization. If the signal contains sufficient high-frequency information, macaque monkeys may detect vertical displacements of only 3° to 4° (Fig. 3.13). This observation corresponds with the expectations based on the cues for perception of elevation discussed above. However, as shown in Figure 3.12, it is likely that the perception of the sound azimuth is more accurate than the perception of elevation for most signals.

Table 3.1 presents the acuity of vertical localization for representative mammals for the best signal tested. With a vertical acuity of 13°, the opposum *Didelphis virqinianus* (Ravizza and Masterton 1972) was the least acute mammal tested, while the bottlenose dolphin *Tursiops truncatus* (Renaud and Popper 1975), at 2°, was the most precise vertical localizer. However, the literature is too sparse to permit much exploration of how the role of pinna shape or size, visual field size, or brainstem anatomy correlates with vertical acuity. Hopefully, investigators in the future will direct further attention to the problem of the perception of elevation.

TABLE 3.1. Vertical localization acuity in representative mammals.

Group	Species	Acuity	Source
Marsupialia	Opossum	13°	Ravizza and Masterton 1972
Carnivora	Cat	4°	Martin and Webster 1987
Primates	Rhesus/Pig-tailed Monkey	3°	Brown et al. 1982
	Human	3°	Wettschurek 1973
Cetacea	Dolphin	2°	Renaud and Popper 1975

The data summarized in this table are rounded to the nearest integer and are for the best signal tested. In some instances, the test signal was a pure tone; in most cases, however, the best test signal was a band of noise, a click, or a species-specific vocalization.

4.6 The Perception of Acoustic Proximity

The perception of acoustic distance, or acoustic proximity, has received very little formal study. Brown and Brown (in preparation) have measured the minimal perceptible change in acoustic distance for human listeners in a forest habitat at a reference distance of 50 m. Using the speech utterance "hey" and a 1-kHz tone for the stimuli, they found that subjects would use changes in loudness, or sound amplitude, if the the magnitude of the stimulus at its source was held constant as the distance was varied. However, if signal amplitude was adjusted to compensate for changes in distance (and if random amplitude fluctuations were introduced), subjects were only able to perceive changes in acoustic proximity for the spectrally complex speech stimulus. This fact indicates that human listeners presumably used changes in sound quality as described in Section 2.6 to detect changes in acoustic distance. Figure 3.14 shows that human listeners could perceive a 10% change in acoustic distance when the source level was fixed for both the tone and speech stimuli. This finding shows that loudness, or auditory image constancy, is an important cue for the perception of changes in acoustic proximity when it is available for processing (the amplitude of the signal is fixed). The detection of a 10% change in acoustic distance in a forested site compares closely with distance discrimination thresholds of about 6% for reference distances of 6 to 49 m on an open athletic field (Strybel and Perrott 1984) and with distance discrimination thresholds of about 6% for reference distances of 1 to 2 m in an anechoic room (Ashmead, LeRoy, and Odom 1990). The scattering of sound in the forested habitat will change the rate of sound attenuation with respect to distance relative to that in open environments (Waser and Brown 1986). Sound propagation is complicated because the elevation of the source and receiver and the frequency of the signal have strong effects. Nevertheless, signals in the speech range, at the elevation of the human head, tend to be propagated better in forested than open habitats

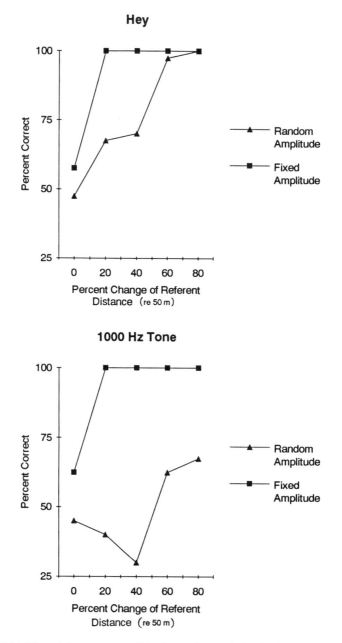

FIGURE 3.14. The minimum perceptible change in proximity for human listeners. The test signals were the word "Hey" (top panel) and a 1000 Hz tone (bottom panel). The reference distance was 50 m. Testing was conducted in a forested habitat. The triangles indicate detection when the intensity of the signal was held constant; the squares indicate detection when the level of the signal was randomized and adjusted to compensate for changes in loudness with distance (see text).

(Waser and Brown 1986; Brown, Gomez, and Waser in preparation). That is, in forested compared to open habitats, a greater change in propagation distance will be required to produce a unit change in the level of the signal, and this likely accounts for the difference in the thresholds reported in open fields and anechoic environments compared to that observed in forested environments.

Under most natural situations, sound amplitude is not the only available cue for the perception of a change in acoustic proximity. When the amplitude of the signal is adjusted to compensate for changes in transmission distance and varies randomly from trial to trial, loudness, or auditory image constancy, is no longer a viable cue. Nevertheless, human subjects were still able to perceive changes in acoustic proximity when tested with a complex speech stimulus. The data in Figure 3.14 show that subjects could perceive a change of 44% of the reference distance under these conditions. In an anechoic room, Ashmead, LeRoy, and Odom (1990) reported that human listeners could detect changes in distance of about 16% at reference distances of 1 to 2 m when the amplitude of the test and reference stimuli were equated. Brown and Brown (in preparation) argued that spectral changes and reverberation

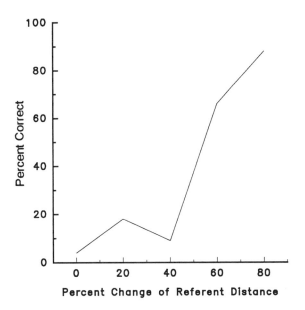

FIGURE 3.15. The minimum perceptible change in proximity in blue monkeys (*Cercopithecus mitis*). The test signal was the pyow vocalization. The reference distance was 50 m. The signal was broadcast and rerecorded at transmission distances of 5, 10, 20, 30, 40, and 50 m. Broadcasts were conducted at an elevation of 7 m in Kibale forest in western Uganda. The amplitude of the test signal was randomized between trials and adjusted to compensate for changes in loudness with distance.

were the most prominent cues underlying the perception of changes of distance.

The ability to perceive changes in sound quality associated with changes in acoustic distance has been measured in blue monkeys (*Cercopithecus mitis*). Figure 3.15 shows that blue monkeys can detect a change in proximity of 54% for the pyow vocalization broadcast in their natural habitat. However, Brown and Brown (in preparation) argue that different acoustic signals and different acoustic habitats, each associated with specific sound propagation characteristics, would strongly influence the ability of listeners to perceive changes in the proximity of acoustic sources. Nevertheless, these observations suggest that reflection of the wave front by tree trunks and other surfaces and frequency-specific attenuation may change or distort acoustic signals in a manner that provides a reliable and perceptually useful index of acoustic distance. It is conceivable that organisms residing in various habitats may have developed signals that are particularly well suited to permit listeners to ascertain the distance to the vocalizer. Furthermore, it is possible that some calls possess an acoustic structure that makes it possible to detect small changes in the proximity of the vocalizer, while other calls may tend to obscure the available distance cues (Brown and Brown in preparation).

5. Conclusion

Mammals have a sense of the azimuth, elevation, and distance of the source of acoustic events. However, the resolution of sound position is not equal in all three coordinates. The available data suggest that for most mammals the acuity of resolution of sound source azimuth is greater than that for elevation, and the acuity of resolution for sound source elevation is greater than that for distance. Hence, the minimal audible change in acoustic locus for azimuth, elevation, and distance may be described by the surface of an ellipsoid, a three-dimensional figure oriented such that the width is less than the height which, in turn, is less than the length. A theoretical three-dimensional minimal perceptible change in locus ellipsoid is illustrated in Figure 3.16.

All three coordinates of sound source localization are important biologically. However, because the cues that underlie the perception of azimuth, elevation, and distance are so dissimilar, it is possible that subjects may experience abnormalities or disorders that impair perception in one dimension, yet leave relatively intact perception in the other two dimensions. Furthermore, it is possible that the ecology and life history of different species have lead to enhanced sensitivity for localization in one coordinate relative to that in another. Terrestrial species may have been selected to maximize acuity for source azimuth, while marine organisms and arboreal species have been selected for enhanced acuity for source elevation, and forest-living species may have been selected for greater acuity for source distance. Because researchers have studied the perception of sound source azimuth and con-

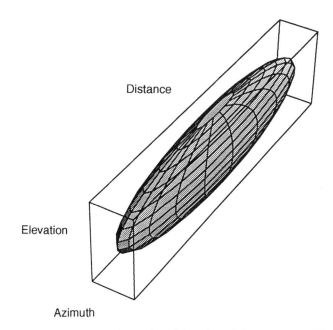

Distance

Elevation

Azimuth

FIGURE 3.16. A theoretical volume describing the minimum perceptible change in locus of a broad bandwidth sound. The reference locus is the center of the ellipsoid and the just perceptible change in locus in any direction is given by the distance from the center to any point on the surface of the volume. In the ellipsoid drawn here, the resolution for changes in azimuth are two times that for changes in elevation and eight times that for changes in distance. The actual dimensions of the volume describing the minimally perceptible change in space would be influenced by the acoustics of the habitat (test environment) and the temporal and spectral complexity of the test signal.

ducted far fewer studies on the localization of source elevation or distance, it is not possible to evaluate differences in ecology or life history and the relative sensitivity of different species to the three coordinates of sound source position.

5.1 Opportunities for Future Research

Greater comparative research is needed in the localization of source elevation and distance. As noted in Section 2.6, it is particularly important to identify the cues employed for the localization of sound source distance under natural conditions. Research in this area is just in its infancy, and different species may use different cues under different situations to estimate the proximity of acoustic events. In addition to the question of acoustic distance, additional research is needed to study the neural mechanisms used in the localization of sound source elevation and distance. In the area of vertical localization, additional research is needed to examine how marine mammals with either no pinna, or a small pinna, are able to localize. It is presently unknown if marine mammals experience asymmetries in elevation-dependent transformation functions for the two ears or if vertical localization is accomplished by some other mechanism.

An additional area for research is the study of disorders of localization. In rats, sound localization abilities appear to decline with age (Brown 1984), and this decline appears to correlate with degeneration in the central auditory system (Keithley and Feldman 1979; Casey and Feldman 1982), rather than with a loss of inner ear hair cells per se. It is possible that disturbances in directional hearing may occur independently of any changes in hearing thresholds measured in the quiet. Additional research is needed to determine if impediments in sound localization abilities are associated with any of the class of difficulties attributed to central auditory processing disorders. Finally, additional research is required to determine how sound localization systems operate in a world where reverberation and masking are prominent. Virtually all sound localization studies are conducted in a free-field setting where reflections and noise are nearly absent (though see Brown 1982b; Brown and May 1990), yet localization mechanisms must normally function under very complicated acoustical conditions. Most older humans reach a point where hearing becomes very effortful in restaurants and other reverberant settings where multiple noise sources are common. It remains unclear how accurate localization is under these conditions for listeners with either normal or impaired hearing. Furthermore, it is unknown if impediments in sound localization mechanisms are associated with hearing difficulties in these settings. It is hoped that the comparative and ecological focus explored by this chapter will encourage researchers to develop new approaches to the study of the spatial dimensions of auditory images, their properties in human and nonhuman auditory perception, and in normal and impaired listeners.

Acknowledgments. I thank F. E. Brown and J. M. Sinnott who commented on an earlier version of this chapter. Portions of the research reviewed in this chapter were supported by National Institutes of Health Grant RO1 DC00164.

References

Anderson DJ (1973) Quantitative model for the effects of stimulus frequency upon synchronization of auditory nerve discharges. J Acoust Soc Am 54:361–364.

Aronson E, Rosenbloom S (1971) Space perception in early infancy: Perception within a common auditory-visual space. Science 172:1161–1163.

Ashmead DH, LeRoy D, Odom RD (1990) Perception of the relative distances of nearby sound sources. Percept Psychophys 47:326–331.

Ayrapet'yants ES, Konstantinov AI (1974) Echolocation in Nature. An English translation of the National Technical Information Service, JPRS 63326-1-2.

Batteau DW, Plante RL, Spencer RH, Lyle WE (1965) Localization of sound: Part 5. Auditory perception. US Navy Ordnance Test Station Report, TP 3109, Part 5.

Beecher MD, Harrison JM (1971) Rapid acquisition of an auditory location discrimination by rats. J Exp Anal Behav 16:193–199.

Beer CG (1969) Laughing gull chicks: Recognition of their parent's voices. Science 166:1030–1032.

Beer CG (1970) Individual recognition of voice in social behavior of birds. In: Lehrman DS, Hinde RA, Shaw E (eds) Advances in the Study of Behavior, Volume 3. New York: Academic Press, pp. 27–74.

Brenowitz EA (1982) The active space of red-winged blackbird song. J Comp Physiol 147:511–522.

Brown CH (1982a) Ventroloquial and locatable vocalizations in birds. Z Tierpsychol. 59:338–350.

Brown CH (1982b) Auditory localization and primate vocal behavior. In: Snowdon CT, Brown CH, Petersen MR (eds) Primate Communication. Cambridge: Cambridge University Press, pp. 144–164.

Brown CH (1984) Directional hearing in aging rats. Exp Aging Res 10:35–38.

Brown CH (1989) The active space of blue monkey and grey-cheeked mangabey vocalizations. Anim Behav 37:1023–1034.

Brown CH, Gomez R (1992) Functional design features in primate vocal signals: The acoustic habitat and sound distortion. In: Nishida T, McGrew WC, Marler P, Pickford M, de Waal FMB (eds) Topics in Primatology, Volume One, Human Origins. Tokyo: University of Tokyo Press, pp. 177–198.

Brown CH, May BJ (1990) Sound Localization and binaural processes. In: Berkley MA, Stebbins WC (eds) Comparative Perception, Volume 1. New York: John Wiley and Sons, pp. 247–284.

Brown CH, Waser PM (1988) Environmental influences on the structure of primate vocalizations. In: Todt D, Geodeking P, Symmes D (eds) Primate Vocal Communication. Berlin: Springer-Verlag, pp. 51–66.

Brown CH, Beecher MD, Moody DB, Stebbins WC (1978a) Localization of pure tones in Old World monkeys. J Acoust Soc Am 63:1484–1494.

Brown CH, Beecher MD, Moody DB, Stebbins WC (1978b) Localization of primate calls by Old World monkeys. Science 201:753–754.

Brown CH, Beecher MD, Moody DB, Stebbins WC (1979) Locatability of vocal signals in Old World monkeys: Design features for the communication of position. J Comp Physiol Psychol 93:806–819.

Brown CH, Beecher MD, Moody DB, Stebbins WC (1980) Localization of noise bands by Old World monkeys. J Acoust Soc Am 68:127–132.

Brown CH, Schessler T, Moody DB, Stebbins WC (1982) Vertical and horizontal sound localization in primates. J Acoust Soc Am 72:1804–1811.

Butler RA (1969) Monaural and binaural localization of noise bursts vertically in the median sagittal plane. J Aud Res 9:230–235.

Cain WS, Marks LE (1971) Stimulus and Sensation. Boston: Little, Brown and Co.

Casey MA, Feldman ML (1982) Aging in the rat medial nucleus of the trapezoid body. Neurobiol Aging 3:187–195.

Casseday JH, Neff WD (1973) Localization of pure tones. J Acoust Soc Am 54:365–372.

Chambers RE (1971) Sound localization in the hedgehog (*Paraechi nus hypomelas*). Unpublished Master's Thesis, Florida State University, Tallahassee.

Clements M, Kelly JB (1978a) Directional responses by kittens to an auditory stimulus. Dev Psychobiol 11:505–511.

Clements M, Kelly JB (1978b) Auditory spatial responses of young guinea pigs (*Cavia porcellus*) during and after ear blocking. J Comp Physiol Psychol 92:34–44.

Coleman PD (1963) An analysis of cues to auditory depth perception in free space. Psychol Bull 60:302–315.

Fay RR (1988) Hearing in Vertebrates: A Psychophysics Databook. Winnetka, IL: Hill-Fay Associates.

Feng AS, Gerhardt HC, Capranica RR (1976) Sound localization behavior of the green tree frog (*Hyla cinerea*) and the barking tree frog (*H. gratiosa*). J Comp Physiol 107:241–252.

Firestone FA (1930) The phase differences and amplitude ratio at the ears due to a source of pure tones. J Acoust Soc Am 2:260–270.

Gamble EA (1909) Intensity as a criterion in estimating the distance of sounds. Psychol Rev 16:416–426.

Gardner MB (1973) Some monaural and binaural facets of median plane localization. J Acoust Soc Am 54:1489–1495.

Gish SL, Morton ES (1981) Structural adaptations to local habitat acoustics in Carolina wren songs. Z Tierpsychol 56: 74–84.

Gottlieb G (1965) Imprinting in relation to parental and species identification by avian neonates. J Comp Physiol Psychol 59:345–356.

Harris JD (1972) A florilegium of experiments on directional hearing. Acta Otolaryngol Suppl 298:1–26.

Harrison JM (1990) Simultaneous auditory discriminations. J Exp Anal Behav 54:45–51.

Harrison JM (1992) Avoiding conflicts between the natural behavior of the animal and the demands of discrimination experiments. J Acoust Soc Am 92:1331–1345.

Harrison JM, Downey P (1970) Intensity changes at the ear as a function of azimuth of a tone: A comparative study. J Acoust Soc Am 56:1509–1518.

Harrison JM, Irving R (1966) Visual and nonvisual auditory systems in mammals. Science 154:738–743.

Harrison JM, Downey P, Segal M, Howe M (1971) Control of responding by location of auditory stimuli: Rapid acquisition in monkey and rat. J Exp Anal Behav 15: 379–386.

Hartley RVL, Fry TC (1921) The binaural localization of pure tones. Phys Rev 18: 431–442.

Hebrank J, Wright D (1974) Spectral cues used in the localization of sound sources in the median plane. J Acoust Soc Am 56:1829–1834.

Heffner HE, Heffner RS (1984) Sound localization in large mammals: Localization of complex sounds by horses. Behav Neurosci 98:541–555.

Heffner HE, Heffner RS (1985) Sound localization in wild Norway rats (*Rattus norvegicus*). Hear Res 19:151–155.

Heffner HE, Masterton RB (1980) Hearing in glires: Domestic rabbit, cotton rat, feral house mouse, and kangaroo rat. J Acoust Soc Am 68:1584–1599.

Heffner RS, Heffner HE (1982) Hearing in the elephant (*Elephas maximus*): Absolute sensitivity, frequency discrimination, and sound localization. J Comp Physiol Psychol 96:926–944.

Heffner RS, Heffner HE (1983) Hearing in large mammals: Horses (*Equus caballus*) and cattle (*Bos taurus*). Behav Neurosci 97:299–309.

Heffner RS, Heffner HE (1985) Auditory localization and visual fields in mammals. Neurosci Abst 11:547.

Heffner RS, Heffner HE (1987) Localization of noise, use of binaural cues, and a description of the superior olivary complex in the smallest carnivore, the least weasel (*Mustela nivalis*). Behav Neurosci 101:701–708.

Heffner RS, Heffner HE (1988a) Sound localization in a predatory rodent, the northern grasshopper mouse (*Onychomys leucogaster*). J Comp Psychol 102:66–71.

Heffner RS, Heffner HE (1988b) Sound localization and the use of binaural cues in the gerbil (*Meriones unguiculatus*). Behav Neurosci 102:422–428.

Heffner RS, Heffner HE (1988c) The relation between vision and sound localization acuity in mammals. Neurosci Abst 14:1096.

Heffner RS, Heffner HE (1988d) Sound localization acuity in the cat: Effect of azimuth, signal duration, and test procedure. Hear Res 36:221–232.

Heffner RS, Heffner HE (1989) Sound localization, use of binaural cues, and the superior olivary complex in pigs. Brain Behav Evol 33:248–258.

Heffner RS, Masterton RB (1990) Sound localization in mammals: Brainstem mechanisms. In: Berkley MA, Stebbins WC (eds) Comparative Perception, Volume 1. New York: John Wiley and Sons, pp. 285–314.

Heffner RS, Richard MM, Heffner HE (1987) Hearing and the auditory brainstem in a fossorial mammal, the pocket gopher. Neurosci Abst 13:546.

Henning GB (1974) Detectability of interaural delay in high-frequency complex waveforms. J Acoust Soc Am 55:84–90.

Henson OW Jr (1974) Comparative anatomy of the middle ear. In: Keidel WD, Neff WD (eds) Handbook of Sensory Physiology: The Auditory System V/1. New York: Springer-Verlag, pp. 39–110.

Hornbostel EM (1923) Beobachtungen über ein-und zweiohrigs Hören. Psychol Forsch 4:64–114.

Houben D, Gourevitch G (1979) Auditory lateralization in monkeys: An examination of two cues serving directional hearing. J Acoust Soc Am 66:1057–1063.

Johnson DH (1980) The relationship between spike rate and synchrony in responses of auditory nerve fibers to single tones. J Acoust Soc Am 68:1115–1122.

Kaufman L (1979) Perception: The World Transformed. New York: Oxford University Press.

Keithley EM, Feldman ML (1979) Spiral ganglion cell counts in an age-graded series of rat cochleas. J Comp Neurol 188:429–442.

Kelly JB (1980) Effects of auditory cortical lesions on sound localization in the rat. J Neurophysiol 44:1161–1174.

Kiang NY-S, Watanabe T, Thomas EC, Clark LF (1965) Discharge Patterns of Single Fibers in the Cat's Auditory Nerve. Cambridge, MA: MIT Press.

King AJ, Hutchings ME, Moore DR, Blakemore C (1988) Developmental plasticity in the visual and auditory representations in the mammalian superior colliculus. Nature 332:73–76.

Klumpp RG, Eady HR (1956) Some measurements of interaural time difference thresholds. J Acoust Soc Am 28:859–860.

Knudsen EI (1983) Early auditory experience aligns the auditory map of space in the optic tectum of the barn owl. Science 222:939–942.

Knudsen EI, Brainard MS (1991) Visual instruction of the neural map of auditory space in the developing optic tectum. Science 253:85–87.

Knudsen EI, Knudsen PF (1985) Vision guides the adjustment of auditory localization in young barn owls. Science 230:545–548.

Knudsen EI, Knudsen PF (1990) Sensitive and critical periods for visual calibration of sound localization by barn owls. J Neurosci 10:222–232.

Knudsen EI, Knudsen PF, Esterly SD (1984) A critical period for the recovery of sound localization accuracy following monaural occlusion in the barn owl. J Neurosci 4:1012–1020.

Knudsen EI, Konishi M (1978) A neural map of auditory space in the owl. Science 200:795–797.

Knudsen EI, Blasdel GG, Konishi M (1979) Sound localization by the barn owl measured with the search coil technique. J Comp Physiol 133:1–11.

Knudsen EI, Esterly SD, du Lac S (1991) Stretched and upside-down maps of auditory space in the optic tectum of blind-reared owls: Acoustic basis and behavioral correlates. J Neurosci 11:1727–1747.

Kuhn GF (1977) Model for the interaural time differences in the azimuthal plane. J Acoust Soc Am 62:157–167.

Kuhn GF (1979) The effect of the human torso, head, and pinna on the azimuthal directivity and on the median plane vertical directivity. J Acoust Soc Am 65:(S1), S8(A).

Kuhn GF (1987) Physical acoustics and measurements pertaining to directional hearing. In: Yost WA, Gourevitch G (eds) Directional Hearing. New York: Academic Press, pp. 3–25.

Linskens HF, Martens MJM, Hendriksen HJGM, Roestenberg-Sinnige AM, Brouwers WAJM, Staak van der ALHC, Strik-Jansen AMJ (1976) The acoustic climate of plant communities. Oecologia (Berlin) 23:165–177.

Makous JC, Middlebrooks JC (1990) Two-dimensional sound localization by human listeners. J Acoust Soc Am 87:2188–2200.

Manley GA (1973) A review of some current concepts of the functional evolution of the ear in terrestrial vertebrates. Evolution 26:608–621.

Martin RL, Webster WR (1987) The auditory spatial acuity of the domestic cat in the interaural horizontal and median vertical planes. Hear Res 30:239–252.

Masterton RB, Heffner HE, Ravizza RJ (1969) The evolution of human hearing. J Acoust Soc Am 45:966–985.

May B, Moody DB, Stebbins WC, Norat MA (1986) Sound localization of frequency-modulated sinusoids by Old World monkeys. J Acoust Soc Am 80:776–782.

McFadden D, Pasanen EG (1976) Lateralization at high frequencies based on interaural time differences. J Acoust Soc Am 59:634–639.

McGregor PK, Krebs JR (1984) Sound degradation as a distance cue in great tit (*Parus major*) song. Behav Ecol Sociobiol 16:49–56.

McGregor PK, Krebs JR, Ratcliffe LM (1983) The reaction of great tits (*Parus major*) to playback of degraded and undegraded songs: The effect of familiarity with the stimulus song type. Auk 100:898–906.

Mendelson MJ, Haith MM (1976) The relation between audition and vision in the human newborn. Monographs of the Society for Research in Child Development, 41, Serial No. 167.

Menzel CR (1980) Head cocking and visual perception in primates. Anim Behav 28:151–159.

Mershon DH, Bowers JN (1979) Absolute and relative cues for the auditory perception of egocentric distance. Perception 8:311–322.

Middlebrooks JC, Green DM (1990) Directional dependence of interaural envelope delays. J Acoust Soc Am 87:2149–2162.

Middlebrooks JC, Makous JC, Green DM (1989) Directional sensitivity of sound-pressure levels in the human ear canal. J Acoust Soc Am 86:89–108.

Mills AW (1958) On the minimum audible angle. J Acoust Soc Am 30:237–246.

Mills AW (1960) Lateralization of high-frequency tones. J Acoust Soc Am 32:132–134.

Muir D, Field J (1979) Newborn infants orient to sounds. Child Dev 50:431–436.

Musicant AD, Chan JCK, Hind JE (1990) Direction-dependent spectral properties of cat external ear: New data and cross-species comparisons. J Acoust Soc Am 87: 757–781.

Norberg RA (1977) Occurrence and independent evolution of bilateral ear asymmetry in owls and implications on owl taxonomy. Philos Trans R Soc London B 282:375–408.

Nyborg W, Mintzer D (1955) Review of sound propagation in the lower atmosphere. US Air Force WADA Tech Rept 54–602.

Payne RS (1962) How the barn owl locates its prey by hearing. Living Bird 1:151–159.

Perrott DR, Ambarsoom H, Tucker J (1987) Changes in head position as a measure of auditory localization performance: Auditory psychomotor coordination under monaural and binaural listening conditions. J Acoust Soc Am 85:2669–2672.

Potash M, Kelly J (1980) Development of directional responses to sounds in the infant rat (*Rattus norvegicus*). J Comp Physiol Psychol 94:864–877.

Pumphery RJ (1940) Hearing in insects. Biol Rev 15:107–132.

Ravizza RJ, Masterton RB (1972) Contribution of neocortex to sound localization in opossum (*Didelphis virginiana*). J Neurophysiol 35:344–356.

Rayleigh JWS (1876) Our perception of the direction of a sound source. Nature (London) 14:32–33.

Rayleigh JWS (1945) The Theory of Sound, Second Edition. New York: Dover Publications.

Renaud DL, Popper AN (1975) Sound localization by the bottlenose porpoise *Tursiops truncatus*. J Exp Biol 63:569–585.

Rheinlaender J, Gerhardt HC, Yager DD, Capranica RR (1979) Accuracy of phonotaxis by the green tree frog (*Hyla cinerea*). J Comp Physiol 133:247–255.

Rose JE, Brugge JF, Anderson DJ, Hind JE (1967) Phase-locked responses to low-frequency tones in single auditory nerve fibers of the squirrel monkey. J Neurophysiol 30:769–793.

Searle CL, Braida LD, Cuddy DR, Davis MF (1975) Binaural pinna disparity: Another auditory localization cue. J Acoust Soc Am 57:448–455.

Shaw EAG (1974) The external ear. In: Keidel WD, Neff WD (eds) Handbook of Sensory Physiology, Vol V/1. Berlin: Springer-Verlag, pp. 455–490.

Simpson WE, Stanton LD (1973) Head movement does not facilitate perception of the distance of a source of sound. Am J Psychol 86:151–159.

Spigelman MN, Bryden MP (1967) Effects of early and late blindness on auditory spatial learning in the rat. Neuropsychologia 5:267–274.

Strybel TZ, Perrott DR (1984) Discrimination of relative distance in the auditory modality: The success and failure of the loudness discrimination hypothesis. J Acoust Soc Am 76:318–320.

Terhune JM (1974) Directional hearing of the harbor seal in air and water. J Acoust Soc Am 56:1862–1865.

von Békésy GV (1938) Über die Entstehung der Entfernungsempfindung beim Hören.

Akust Z 3:21–31. (Available in English in Wever EG (ed) Experiments in Hearing. New York: John Wiley and Sons, 1960, pp. 301–313.

Waser PM (1977) Sound localization by monkeys: A field experiment. Behav Ecol Sociobiol 2:427–431.

Waser PM, Brown CH (1984) Is there a sound window for primate communication? Behav Ecol Sociobiol 15:73–76.

Waser PM, Brown CH (1986) Habitat acoustics and primate communication. Am J Primatol 10:135–154.

Wertheimer M (1961) Psychomotor coordination of auditory and visual space at birth. Science 134:1692.

Wettschurek RG (1973) Die absoluten Unterschiedswellen der Richtungswahrnehmung in der Medianebene beim natürlichen Hören, sowie beim Hören über ein Kunstkopf-Übertragungssystem. Acoustica 28:197–208.

Whittington DA, Hepp-Reymond MC, Flood W (1981) Eye and head movements to auditory targets. Exp Brain Res 41:358–363.

Whitehead JM (1987) Vocally mediated reciprocity between neighboring groups of mantled howling monkeys, *Aloutta palliata palliata*. Anim Behav 35:1615–1627.

Wightman FL, Kistler DJ (1989a) Headphone simulation of freefield listening. I: Stimulus synthesis. J Acoust Soc Am 85:858–867.

Wightman FL, Kistler DJ (1989b) Headphone simulation of freefield listening. II: Psychophysical validation. J Acoust Soc Am 85:868–878.

Wightman FL, Kistler DJ (1992) The dominant role of lowfrequency interaural time differences in sound localization. J Acoust Soc Am 91:1648–1661.

Wiley RH, Richards DG (1978) Physical constraints on acoustic communication in the atmosphere: Implications for the evolution of animal vocalization. Behav Ecol Sociobiol 3:69–94.

Woodworth RS (1938) Experimental Psychology. New York: Holt.

4
How Monkeys Hear the World: Auditory Perception in Nonhuman Primates

William C. Stebbins and David B. Moody

1. Introduction

That nonhuman primates, and, for that matter, other mammals, have an effective and well-adapted ear on their world as we do on ours is unmistakably true. The once arguable question of whether that is knowable in animals without benefit of language is no longer so. In fact, other animals can be questioned about their sensory experience and about their ability to discriminate the finest details of the auditory stimuli that impact on their lives. Methods exist to investigate their perception of the complex auditory patterns that foretell of the wing beat or footfall of an advancing predator, the departure of a selected quarry, or the vocal communicative signal of a conspecific. Although field observations are heavily relied on to determine what is important in the natural lives and auditory experience of these animals, the specific questions regarding their sensory acuity or perceptual judgment are necessarily framed in the laboratory environment where control over the relevant variables allows measurement of sensory or perceptual function in as precise and reliable a manner as similar experiments with humans.

An animal's behavior in response to simple or complex auditory signals can be described in a series of laboratory experiments. It is this relationship between signal and behavior that defines sensory detection, discrimination, or even perceptual judgment. A stimulus responded to 50% of the time, for example, may be interpreted as an estimate of that animal's absolute threshold or minimum detectable level of stimulation. Similarly, the least difference between two levels or frequencies of stimulation responded to half the time is the difference threshold or an estimate of the minimum resolvable difference between two stimuli. These and other measures reflect the auditory acuity of an animal or the limits of its sensory resolving power and are, for the most part, determined by its basic peripheral physiological equipment. More frequently, in their daily lives, animals function well within those sensory limits and may often respond similarly to stimuli that they could distinguish if forced to do so. For example, vervet monkey (*Cercopithecus aethiops*) alarm calls in response to an avian predator may differ in their acoustic structure depending on the individual caller or they may vary in the same caller depending on the precise circumstances (Seyfarth, Cheney, and Marler 1980a,b). Yet another vervet may respond in the same way to such calls despite this acoustic diversity. Although well within its power, it may not be to the listener's advantage to respond differentially, e.g., to take stock of the individual caller's identity before taking evasive action. Classification or categorization is an important aspect of perception that is less clearly related to the underlying peripheral physiology and to the antecedent stimuli. Its provenance may reside in learning or development or it may be present at birth and central neural mechanisms are apt to exert a more significant influence (see Heffner and Heffner 1990). But, in the same way that an animal's auditory acuity is measured, the nature of its classificatory skills within a discrimination paradigm is studied. An animal responds similarly to stimuli within a given category and differently (discrimination) to stimuli outside that category. The important procedural differences between the study of acuity and of perceptual judgment or classification will be described in Section 3.

We will review here the literature on auditory perception in nonhuman primates. Perception is used in its generic sense to include both measures of auditory acuity for simple stimuli, such as pure tones, and the classification and categorization of more complex auditory signals, such as those used in vocal communication. Communication, of course, includes both production and perception. Although we will treat primarily the perceptive side of primate communication, it is not possible to avoid some reference to primate vocal production since, of course, the sender is the signal source. The approach is comparative and we make some comparisons to other animals (see also Long, Chapter 2; Brown, Chapter 3). We will also try to say something about primate auditory perception in an evolutionary context (see also Stebbins and Sommers 1992).

Despite the fact that laboratory studies represent the core of information in this chapter, they are not entirely separable from the observational studies on communication that are carried out in the field and provide the necessary ecological context for the more analytical laboratory experiments. Stimuli are taken directly from the field for experiments on the perception of communication sounds. "Complex," as in complex auditory signals, is most often a euphemism for "biologically relevant." In the process of asking significant questions about perception in the laboratory, researchers must rely heavily on the social context in which those signals occur. The reasons should become more apparent as these studies are presented in Section 3.3. But even the more basic experiments on the detection and discrimination of simple stimuli must refer to the natural context or setting. Auditory thresholds for pure tones determined in an utterly silent room where the only natural object is the animal subject must be recalibrated for the natural habitat where listening conditions are seldom so ideal. The field primatologist can make only the crudest judgment about an animal's sensory acuity, and yet this kind of information is often essential for the inferences that must be made and the conclusions that must be drawn from the behavior observed in the natural setting.

For example, Donisthorpe (1958) noted how very close she was able to approach a gorilla (*Gorilla gorilla*) troop without apparently unsettling its members. She inferred, on the basis of this observation, that gorillas are not especially sensitive to sound, at least not as sensitive as humans. Five years later, Schaller (1963) was to dispute this interpretation of Donisthorpe's observations, arguing, from his own view of the distance, that gorillas can be disturbed by strange, often man-made sounds and that the auditory sensitivity of gorillas is comparable to humans. Was Donisthorpe approaching the gorillas more quietly than Schaller? Were the gorillas intentionally ignoring her? Was Schaller dressed in bright colors or was he upwind of the gorilla troop? These questions will never be answered. Anecdotal evidence is, of course, the stuff of which early comparative psychology was made and convinced us that horses can compute, dogs are color-blind, and fish and snakes are deaf. None of the above are true. The moral, of course, is that perceptual questions are best studied in the laboratory, where an animal can be removed from its rich environmental milieu, where stimuli can be controlled and measured with the necessary precision, and where the sensory system of interest can be studied behaviorally in isolation, if necessary, from other sensory systems.

It has been argued that the laboratory environment is an artificial one, dissimilar from the real world in all of its characteristics. How a gorilla responds under natural conditions will bear little resemblance to its reaction in the laboratory. We have no cavil with this objection if it refers to foraging or social behavior, for such behaviors are unsuitable for laboratory study. But if there is a unique interest in an animal's resolution of sensory stimulation or in its perceptual evaluation of such information from its habitat, then there is little choice but to resort to the laboratory for an answer. Examples of the former are the limits of an animal's auditory acuity and the range of stimulation over which it can operate (for an earlier review, see Stebbins 1971). And of the latter, within that range, it might be asked how an animal chooses, classifies, or categorizes those signals that it must engage. Or, put somewhat differently, what is an animal capable of out at the margin, and what does it do ordinarily when not forced to function at the limits of its capacity? Both questions require an experimental paradigm in which the animal is instructed to choose among alternatives. The answers to these questions and the various approaches taken to obtain them from a limited variety of nonhuman primates form the subject matter of this chapter.

2. Auditory Acuity

2.1 *Audibility*

Perhaps the most fundamental problem for perception is the choice between the presence or absence of a signal. On those occasions when the signaler is at some distance from the receiver, the light is dim, the ambient noise level is high, the air is turbulent, and the context is one of eat or be eaten, mate or lose reproductive advantage, the receiver must call upon the full extent of its capabilities. If an animal's acuity in the laboratory can be measured, then it can be predicted with reasonable accuracy what the animal can do in its own niche, at the margin, given the context and given the signal to which it is called upon to respond. In the experiment, the acoustic signal becomes less and less intense until the subject fails to discriminate between sound and silence, and we speak of an absolute threshold for sound and characterize the animal's audibility thus. Such an experiment was completed on the chimpanzee (*Pan paniscus*) in the 1930s (Elder 1933, 1934). Elder reported that this primate hears much as humans do, although it has a higher upper frequency limit (30 kHz as opposed to 20 kHz for humans; Elder 1935; see also Kojima 1990). These results are currently the best estimate for the gorilla, who has never been enticed into the laboratory for such an experiment. Schaller (1963) may have been closer to reality than Donisthorpe (1958), but his was at best an educated guess and sufficiently imprecise to be of little value as a scientific datum.

Elder's (1933, 1934, 1935) experiments on the chimpanzee and the preliminary findings of Wendt (1934) on several species of Old and New World monkeys represented the first serious and systematic attempt to measure audibility in a nonhuman primate. In spite of the lack of an adequate acoustic and behavioral technology, these early experiments represented a tour de force, and the findings appear reliable and valid by recent standards. An important distinction between human and animal psychophysics is the existence of a language barrier between animal subject and human experimenter. The questions must be couched in terms that the animal can understand and framed with sufficient precision that the answer will be both valid and reliable. For this purpose, the technology of operant conditioning with positive reinforcement is used to train the animal as an observer of sensory events (see Stebbins 1970; Moody, Beecher, and Stebbins 1976; Stebbins, Brown, and Petersen 1984; R. Heffner and Masterton 1990). Methods in different laboratories differ in some details, but uniform results from these laboratories indicate that these methodological differences are not important (R. Heffner and Masterton 1990, p. 287).

We describe briefly one such procedure for measuring absolute auditory thresholds (Stebbins, Brown, and Petersen 1984). With minor variations, it is the one that we have used throughout in our studies of the auditory perception of nonhuman primates. Within a soundproofed acoustic chamber, the animal, seated in a special restraint chair, is trained with food as a reinforcer to listen and then respond to tonal stimuli varying in level and frequency. Earphones permit monaural testing. The animal responds to a visual "ready" signal by making manual contact with a metal tube and reports an auditory signal by breaking contact with the tube. The error rate is kept low by postponing the opportunity for reinforcement when errors occur, and guessing is monitored by the use of "catch trials" which are regularly programmed trials

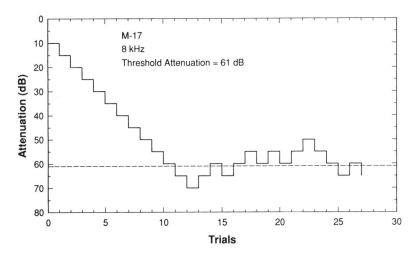

FIGURE 4.1. The process of determining auditory threshold for one monkey (M-17) at 8 kHz by the staircase or tracking method. Correct detections cause the tone to be attenuated, whereas failure to hear is followed by an increase in tone intensity on the next trial. Threshold, calculated as the average transition value from the function, is indicated by the horizontal dotted line. (From Stebbins 1973.)

but without stimuli. When the subject has learned to respond quickly, accurately, and decisively to clearly audible tones when they are presented, a threshold testing procedure is instituted. Each time the animal responds correctly to the tone, its level is lowered for the next presentation; when the animal fails to respond, the level is subsequently increased. The animal's threshold for minimum audibility is then calculated (in the same way it would be for a human subject) as that level of the tone to which it responds correctly half the time.

The typical up-down or staircase pattern characteristic of a Békésy audiogram but taken from a macaque monkey (*Macaca nemestrina*) is presented in Figure 4.1; the calculated threshold is shown by the dashed line. These thresholds then determined at several pure-tone frequencies are put together as a composite audibility function that reflects the range of hearing of the subject species and its absolute sensitivity within that range. An example for a macaque monkey is shown in Figure 4.2. It can then be compared with threshold functions or audiograms obtained from several different primate species and one nonprimate mammalian species, the tree shrew (*Tupaia glis*), seen in Figure 4.3. From these findings, several comparative observations can be made (see also Masterton, Heffner, and Ravizza 1969). First, the tree shrew and the one prosimian shown here (the bush baby *Galago senegalensis*) have a frequency range of hearing that extends to about 60 kHz, which is characteristic of many of the nonprimate mammals. They also appear somewhat less sensitive than the other primates at the lower frequencies.

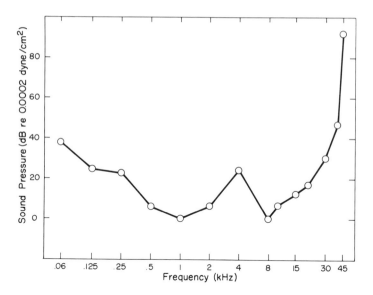

FIGURE 4.2. Threshold of hearing function for one monkey (*Macaca nemestrina*). (From Stebbins 1978.)

Humans, at the other extreme, have the most limited frequency range of hearing among the primates, and, in fact, among the mammals as a group, but are as sensitive as the other primates in the midrange frequencies. The macaques and most other Old World monkeys are intermediate. Their range of hearing extends half an octave above that of humans, to about 30 or 45 kHz. Limited findings from a few species of New World monkeys indicate that their audibility is comparable to that of their Old World cousins (Beecher 1974a,b; Green 1975a).

Generalizations for the major primate groups from the data presented above are made more difficult by the exceptions found within the various taxa. For example, there are at least two species of prosimians (*Nycticebus* and *Perodicticus*, see H. Heffner and Masterton 1970) whose hearing range resembles that of the Anthropoid primates (Old and New World monkeys). Among Old World monkeys, there are one or two species that show unusual acuity within their audible range at frequencies that are contained in their vocalizations (Brown and Waser 1984). These differences are large enough (18 dB) to be considered the consequence of specific pressures related to communication in these animals' habitat. Perhaps the only safe generalizations are, first, that the Anthropoid primates as a group have a more restricted range of hearing than the other mammals, although they may be more sensitive than many other mammals to frequencies below about 8 kHz. Second, humans, with an audibility range extending only to 20 kHz, appear to represent the extreme among primates and among all mammals tested,

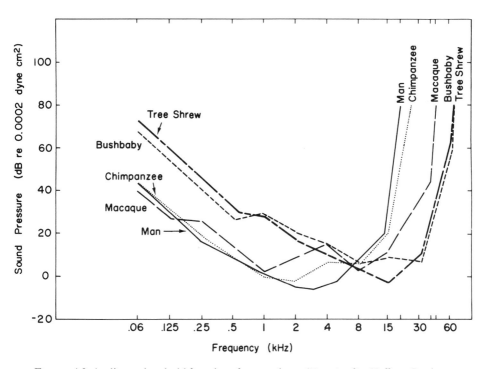

FIGURE 4.3. Auditory threshold functions for tree shrew (*Tupaia glis*; Heffner, Ravizza, and Masterton 1969a); bush baby (*Galago senegalensis*; Heffner, Ravizza, and Masterton 1969b); macaque (*Macaca nemestrina*; Stebbins, Green, and Miller 1966); chimpanzee (*Pan paniscus*; Elder 1934; Farrer and Prim 1965); and man (*Homo sapiens*; Sivian and White 1933). The tree shrew is an insectivore; the bush baby is a prosimian primate; and the macaque monkey, chimpanzee, and man are Anthropoid primates. (From Stebbins 1971.)

with only the rare exception (the chinchilla has a similar audiogram; Clark et al. 1974; see also Long, Chapter 2).

Over much of its range, the Old World monkey's audibility function resembles the human's closely enough such that if Donisthorpe (1958) and Schaller (1963) had been watching these animals rather than gorillas, their own hearing, if normal, would have provided a good indication of what this monkey could hear. With caution, Elder's findings on the chimpanzee might be extrapolated to the gorilla to suggest that the gorilla also hears as humans do. But knowledge of an animal's threshold of hearing is only the first step in characterizing that animal as a receiver of acoustic information from the environment. Audible sounds differ in many characteristics. Frequency difference is related to what humans understand as pitch, and the minimum perceptible frequency difference for an animal may tell something about how meaning is conveyed to that animal by changes in frequency or how kin or individual recognition occur in the natural habitat. Level differences in the signal are directly related to loudness perception and can indicate the degree of affect or motivation in the sender. Both level and frequency differences provide important cues to an animal in locating the source of a signal. The ability to gauge the distance accurately and locate the whereabouts of another animal on the basis of acoustic cues provides an advantage for vertebrates in finding mates or prey and in avoiding predators (see Brown, Chapter 3).

2.2 Pure-Tone Frequency Discrimination

Frequency difference thresholds for pure tones have been obtained in many species (see Fay 1988). The procedure that we have used in macaques is a slight variant of the one described in Section 2.1 for absolute thresholds. When the subject first responds by contacting the tube, a repeating tone is presented, followed after a varying interval of time by a second tone differing from the first only in frequency. The animal is then reinforced for releasing the tube as a consequence of the change in frequency. Thresholds are mea-

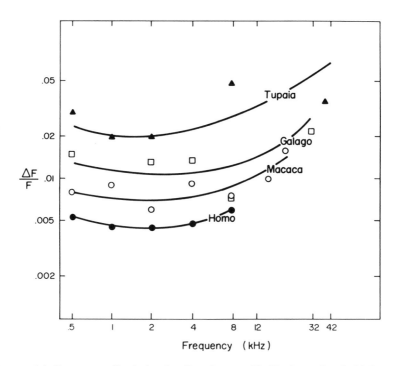

FIGURE 4.4. Frequency discrimination functions at 40 dB above threshold for man and several species of nonhuman primates: *Tupaia glis* (Heffner, Ravizza, and Masterton 1969a); *Galago senegalensis* (Heffner, Ravizza, and Masterton 1969b); *Macaca nemestrina* (Stebbins 1971); and *Homo sapiens* (Filling 1958). (From Stebbins 1971.)

sured by bringing the frequencies closer together and determining the frequency difference (ΔF) to which the animal responds half the time. The problems arise in the training time required to obtain stable thresholds and in the slow and sometimes erratic drift of the thresholds downward over a period of time that is measured in months and even years (see Prosen et al. 1990 for a review and discussion of the problem).

Examples of frequency difference thresholds for several primate species and one nonprimate mammal are shown in Figure 4.4. Here the data are presented in the form of Weber's fraction, $\Delta F/F$, as a function of the standard frequency, F. Note in Figure 4.5 the difference thresholds for frequency, plotted here simply as ΔF across frequency, from several experiments on Old World monkeys and on man. The results of two experiments on Old World monkeys carried out in our laboratory and separated by more than a decade show a noticeable difference in thresholds, although the shape of the functions is the same (see Prosen et al. 1990). The findings from the several human experiments also differ considerably. It is argued that the discrepancy in the results from the different human experiments may be the result of extensive training and possibly subject selection; that claim cannot be made for the differences seen in the monkey experiments.

It is legend in animal psychophysics that all animals (humans included) find the discrimination of small differences in frequency a formidable task—

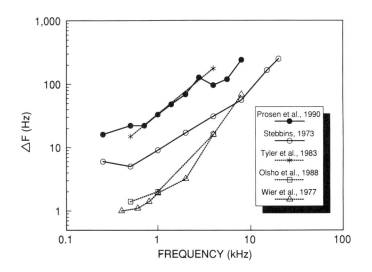

FIGURE 4.5. Difference thresholds for frequency, in Hz, plotted as a function of frequency, determined for the 11 monkeys from the Prosen et al. (1990) study; 3 monkeys reported by Stebbins (1973); unselected, untrained humans (Tyler, Wood, and Fernandes 1983); unselected, well-trained humans (Olsho, Koch, and Carter 1988); and preselected, well-trained humans (Wier, Jesteadt, and Green 1977). (From Prosen et al. 1990.)

far more so, for example, than tone-on-tone masking, the discrimination of complex signals used in speech or in animal communication, or even the discrimination of frequency modulation. Stable baselines are achieved only after months, or even years, of training (Prosen et al. 1990). It seems clear that inadequate stability criteria may produce ambiguous findings, although this is an unlikely explanation for the findings that Prosen el al. present because of the very lengthy periods during which the animals were run in the course of the experiment. There is considerable variability in these measures, as is evident in Figure 4.5.

Sinnott and her colleagues (Sinnott, Petersen, and Hopp 1985; Sinnott, Owren, and Petersen 1987) have pursued this difficult problem in both human and nonhuman primates, and their findings, while interesting, seem to raise more questions than they answer. Their results indicate that frequency difference thresholds in Cercopithicus and macaque monkeys obtained at a standard frequency of 1 or 2 kHz differ depending on whether the comparison frequency is varied in an upward or in a downward direction from the standard. Furthermore, in some species they report that a decrease in frequency results in a lower difference threshold, while in others they suggest that an increase may produce that result. Unfortunately, there are individual exceptions to the rule in both instances where the sample size is only three or four and the individual variance is not presented (Sinnott, Owren, and Petersen 1987). A more complete and fine-grain analysis will be necessary before some of these differences can be judged as either real or important to the species. It is possible that the stability criteria which Sinnott and colleagues applied in training their animals are not sufficiently stringent and that the animals have not yet reached a stable baseline. Some of the findings of Prosen et al. (1990) on the change in frequency difference thresholds over very long time periods would indicate that this might be the explanation. It should be noted that the frequency difference thresholds for those two frequencies employed in the Sinnott, Owren, and Petersen (1987) and Sinnott, Petersen, and Hopp (1985) studies are in the range of those reported by Prosen et al. (1990). On the other hand, the selection of the data presented by Sinnott, Owren, and Petersen (1987) and Sinnott, Petersen, and Hopp (1985) as the "mean of the three lowest difference thresholds obtained over all test sessions" would, we think, given the variability encountered, result in lower thresholds in a relatively short period of training.

Sinnott, Owren, and Petersen (1987) have suggested that greater sensitivity to one direction of frequency change as opposed to another may be related to a similar direction of frequency change in the species communication sounds. This seems unlikely for two reasons. First, the changes in frequency in their experiment were stepwise, discrete frequency changes, while the communication signals to which they referred were frequency modulated—that is, the changes in frequency were continuous. Second, in many of these signals, for example, the rhesus monkey (*Macaca mulatta*) screams, the frequency changes that take place were well above the animal's difference threshold for frequency. However, it is likely that the differences between human and nonhuman primate frequency difference thresholds (Sinnott, Owren, and Petersen 1987; Prosen et al. 1990) are real, and these may be a consequence of the selective pressures related to that most complex form of animal communication—human language. This, too, is a risky assumption for two reasons: first, the anatomical and physiological substrates for frequency difference thresholds are not known with certainty; and, second, the time course of these changes seems comparatively short given the evidence (admittedly shaky) on the earliest origins of human language.

2.3 Frequency Modulation

If the experimental question calls for the detection of frequency modulation (FM) from a pure tone standard rather than one pure tone from another, the context more closely approaches the kind of signal that many birds and mammals, including humans, employ in vocal communication. Over a range of frequencies that spans the frequency range of the vocal communication system of Anthropoid primates, FM detection thresholds are higher than frequency difference thresholds, although both increase with frequency. The findings in Figure 4.6 show that macaques are less sensitive than humans for both frequency difference for pure tones and FM detection thresholds, yet the shape of the functions are alike for man and monkey (Moody and Stebbins 1989).

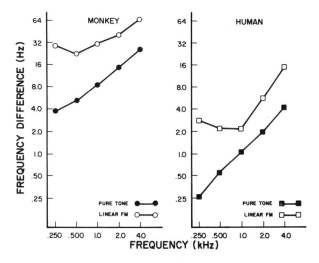

FIGURE 4.6. Frequency modulation (FM) and frequency difference thresholds at various standard frequencies for macaque monkeys and humans. The monkey FM data are from Moody et al. (1986); the monkey frequency difference thresholds are from Stebbins (1975); the human FM data are from Arlinger et al. (1977); and the human frequency difference thresholds are from Nordmark (1968). (From Moody et al. 1986.)

2.4 Frequency Selectivity

While frequency difference thresholds provide an estimate of an animal's ability to resolve the disparity in frequency between two successively presented signals, frequency selectivity refers to the capacity to cull out or separate one frequency from a tonal complex. In the simplest case, one pure tone is masked by another and the threshold is that level of the masker at which the animal can just detect the test tone in its presence. In determining the thresholds, the masking tone is varied in level and frequency while the test or probe tone remains set at one frequency, usually about 10 dB above its threshold previously determined in the absence of masking. In the process of varying the masker, a psychophysical tuning curve is generated. The frequency of the test tone is then changed and the entire process is repeated. Several of these repetitions then yield a family of psychophysical tuning curves, as shown in Figure 4.7. It is the shape of the individual curves that defines the selectivity of the auditory system at each frequency, and it is clear that frequency selectivity, as indicated by the width of the individual functions, is somewhat greater at the higher frequencies.

The common quantitative measure of tuning curve shape is Q_{10}, which provides a relative indication of the sharpness of tuning (selectivity) of a filter and is defined as the center frequency divided by the bandwidth 10 dB above threshold at the center frequency. Unfortunately, Q_{10} has not proven to

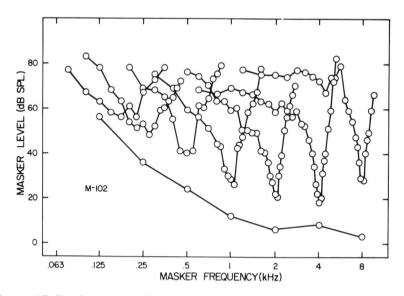

FIGURE 4.7. Simultaneous masking psychophysical tuning curves for subject M-102 (*Macaca nemestrina*). Each point is the median of five separate tests at that frequency. The lower curve indicates absolute threshold at each of the corresponding probe frequencies. (From Serafin, Moody, and Stebbins 1982.)

be a very discriminating measure since it fails to take into account many commonly observed features of tuning curve shape such as the long low-frequency tail. Tuning curves from macaque (Serafin, Moody, and Stebbins 1982) and patas monkeys (*Erythrocebus patas*; Smith, Moody, and Stebbins 1987) are very similar in shape to those obtained from human primates and from other mammals, as are the Q_{10} (see Fay 1988). It is clear, however, that psychophysical tuning curves closely resemble tuning curves for single auditory nerve fibers, and it is likely that these behavioral functions are largely determined by processes that take place at the auditory periphery (Smith, Moody, and Stebbins 1990). As to what a highly frequency-selective animal gains from its environment, one can point to those situations where the extraction of certain frequencies from a complex signal provides an advantage, as in species communication and source identification.

2.5 Pure Tone Intensity Discrimination

Thresholds for level differences in pure tones have been measured in macaque and Cercopithecus monkeys as well as in man (Stebbins 1971,1978; Sinnott, Petersen, and Hopp 1985). The procedure is the same as that employed for pure tone frequency discrimination thresholds. Subjects are trained to detect a level change in a repeating series of tonal stimuli; the level difference is then decreased and thresholds are determined. Difference thresholds for the monkeys are about 2 dB when the stimuli are well above the sound threshold (50 to 60 dB), and do not vary significantly over a frequency range of 500 Hz to 32 kHz (Stebbins 1978). Reported human thresholds are about 1 dB (Jesteadt, Wier, and Green 1977; Sinnott, Petersen, and Hopp 1985) or slightly less (Riesz 1928). Sinnott, Petersen, and Hopp (1985) have reported difficulty in training their monkeys to detect intensity decrements and reported somewhat higher thresholds for decrements as opposed to increments. Whether this is some form of training artifact or a fundamental difference has not yet been confirmed. Level differences are critical in sound localization and in the discrimination of communication sounds.

2.6 Sound Localization

Brown (Chapter 3) has presented a thorough review of sound localization in animals (see also Brown and May 1990; R. Heffner and Masterton 1990). We will therefore deal with localization only very briefly here. Sound localization has been studied systematically in the macaque monkey by procedures very similar to those used for frequency and intensity discrimination. The animal is carefully positioned in a restraint chair in an anechoic chamber. A brief acoustic stimulus is presented repetitively from a speaker at zero degrees azimuth in the horizontal plane. After a varying period of time, the stimulus is presented from a speaker in a different horizontal location and the subject is reinforced for responding to the change in location. Threshold (often referred to as the minimum audible angle) is specified as that angle in degrees of horizontal azimuth at which the animal responds to the stimulus on half the trials. Under optimal conditions, the minimal audible angle for the macaque is about three degrees of arc compared to about one degree for human subjects (Brown and May 1990).

A noisy stimulus is most easily and accurately located. Brown et al. (1980; see also Brown 1976) have shown, for example, that in macaques the minimum audible angle decreases with the frequency bandwidth of the signal. The findings, which are shown in Figure 4.8, provide experimental support for Marler's suggestion that the acoustic features of a signal are the key to the accuracy with which it may be localized (Marler 1955, 1957). It is a successful form of adaptation in some animals, whose alarm calls are highly tonal, that they can warn next of kin and yet escape detection by a nearby predator. From their extensive research on sound localization in many mammalian species, R. Heffner and Masterton (1990) have argued that sound localization

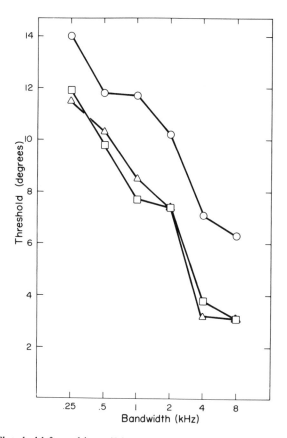

FIGURE 4.8. Threshold for subjects (*Macaca nemestrina*) as a function of bandwidth for stimuli centered at 8 kHz. (From Brown 1976.)

acuity and head size (more accurately, interaural distance) are directly related. There are a few species that are exceptions to this rule, which has led R. Heffner and Masterton to argue that predator species, in general, localize more accurately than prey species and that certain habitats, such as those underground, exercise little, if any, selective pressure for accurate sound localization (R. Heffner and Masterton 1990). Furthermore, the precise localization of sound permits an animal to employ its visual system for a closer examination of the object of interest, thus perhaps relieving some of the pressure for accuracy on the auditory system (see R. Heffner and Masterton 1990). Harrison and his colleagues have shown that squirrel monkeys (*Saimiri sciureus*), as well as rats, are extremely sensitive to the location of an acoustic signal and require little, if any, learning to direct their attention to a salient acoustic event (Harrison et al. 1971).

2.7 The Psychological Attributes of an Acoustic Stimulus

The judgment of the "quality" of a stimulus, such as its loudness, brightness, or pitch, presents an interesting problem to the animal psychophysicist. Human subjects are routinely asked to make these judgments, which are often difficult; they are referred to as "subjective," which presumably renders them of questionable validity as scientific data. Yet the findings have proven useful in a variety of situations such as telephonic communication or listening to recorded music. It is the instructions to the subject that provide the real impasse in animal experiments. By contrast, in threshold experiments, the conditions of reinforcement inform the animal of the nature of the discrimination. For example, reinforcement may be given for responses to a tone but not for responding to its absence, or for responding to stimuli that differ on some physical dimension but not when they are alike on that same dimension. However, no such advance specification of the reinforcement contingencies are possible in an experiment on loudness perception, for example. There is no a priori knowledge of what might represent equal loudness for a nonhuman subject; in fact, it is the very question being asked. Thus, techniques very different from the ones described above must be devised to circumvent the language barrier between human experimenter and animal subject.

One such procedure takes advantage of the relationship between the intensity of a stimulus and the latency of a behavioral response. As the intensity of the acoustic stimulus (a pure tone) increases, the latency of the behavioral response (releasing a telegraph key) decreases in an orderly manner, providing a measure of the effect of stimulus level. Reaction time-intensity functions are obtained over the animal's audible frequency range. Equal reaction times at different pure tone frequencies are then taken as representing measures of equal sensory effect, and equal loudness contours are constructed across frequency and bear the same relation to the pure tone threshold function in nonhuman primates as they do in man. For both human and nonhuman primates, higher stimulus levels are required at the upper and lower ends of the audible frequency range in order to be judged equally loud to the mid-range frequencies where sensitivity is the greatest (Stebbins 1966; Pfingst et al. 1975; Kojima 1990). These findings must be taken into account in evaluating effective communication and perception in the natural habitat, along with the many habitat conditions such as the vegetation, air turbulence, and masking effects of other habitat sounds.

3. The Perception of Complex Signals

3.1 Approaches to the Problem

Classical psychophysics on the detection and discrimination of tonal stimuli has provided our current knowledge and characterization of the outer limits of the primate auditory system. Our ever-increasing understanding of auditory physiology and anatomy has suggested some of the basic mechanisms that are responsible for setting these limits. These psychophysical data also begin to give a better idea of what kinds of informational transactions are possible for these animals in their natural setting and of the selective pressures that might have led to their successful adaptation to their acoustic environment. While field studies conducted in the animal's habitat supply important hypotheses on what these animals may perceive under natural conditions, they lack the necessary control that can be found only in laboratory studies. Now that there is some comprehension of what these outer limits are on the processing of acoustic events by the auditory system, it is possible to look more closely at how the system works within those limits and how it processes the complex signals that are typically encountered by an animal in continual interaction with its environment. These include, but are not limited to, the communication sounds of conspecifics. Here, too, as outlined in Section 3.2, 3.3, the coordination of laboratory and ethological or field studies is essential.

3.2 Field Studies

Observations of the social behavior of animals in their habitat have been enhanced greatly in recent years both by increased attention to the possibility of more sophisticated experimentation under natural conditions and the considerable improvements in acoustic technology, particularly in the area of acoustic recording and playback of the communication sounds of conspecifics (Seyfarth, Cheney, and Marler 1980a,b; S. Gouzoules, H. Gouzoules, and Marler 1984; H. Gouzoules, S. Gouzoules, and Marler 1985; Gouzoules and Gouzoules 1989; Cheney and Seyfarth 1990). In many instances, the playback permits the investigator to observe the animal's response to a signal that is under the investigator's control and that the investigator has initiated in a chosen context and without the usual visual and olfactory accompaniment. This enables the field investigator to at least draw some preliminary conclusions about the perception of communication sounds which are of considerable help and a useful guide to the laboratory experimenter (see Section 3.3).

In many ways, Green's research with the Japanese monkey (*Macaca fuscata*) represented a turning point in the field studies of communication and social behavior in nonhuman primates (Green 1975b). For perhaps the first time, there began to be seen the possibility that field studies of animal communication could make a very valuable contribution to the study of animal perception by a substantial increase in the rigor with which they were carried out. Exceptionally precise field recordings of Japanese monkey calls, combined with thousands of hours of careful observation, enabled Green to describe accurately the relationship between call and social context and to draw reasonable conclusions regarding call meaning. Green was able to relate the actual acoustic structure of the call to its social and communicative function. This accomplishment was made possible by the identification of the senders and receivers of individual calls. Previously the receiver had been given too little attention, and the perception of these biologically relevant signals had, for the most part, been only the object of surmise.

For the greater part of his research, Green (1975b) focused on the affiliative contact or "coo" call that came in many different varieties, each with a different message. One of these, the "smooth high coo" was named for the frequency modulation in the call, with peak frequency occurring either early or late in the call. The call with the early peak is most often employed with young animals at some distance from their mother and serves to establish contact; the smooth late high, on the other hand, is usually a sexual solicitation by an estrous female directed toward potential male consorts. Were these calls that appear to serve such different functions clearly discriminable to the animal? In spite of considerable individual variation in call structure, are there features common to each call class that permit unmistakable classification of each call by the listener? Questions such as these regarding the nature of perceptual function were directed to the laboratory for an answer.

A major premise of early students of animal communication was that animal signals were essentially affective, reflecting an animal's momentary drive state and exhibiting little of the cognitive structure which is such a predominant feature of human language (Griffin 1985). This interpretation implied a significant discontinuity, even a qualitative difference, between animal communication and human language. While there is no question that animal signals do function in an affective manner, more recent findings suggest that there is a clear cognitive component in animal communication, and even signals that denote strong emotion in crisis situations, such as aggressive encounters between animals, often carry important additional information. These messages convey far more than the emotional or motivational state of the signaler. In other words, the cognitive element seems to be coupled with the affective one. For example, certain rhesus monkey screams during combat, in addition to indicating the caller's level of arousal, may convey a message that identifies the caller and his relationship to the listener and may therefore be considered a call for help from close relatives (S. Gouzoules, H. Gouzoules, and Marler 1984). A careful acoustic analysis of rhesus monkey screams provided unmistakable documentation of acoustic features in the calls for the potential identification of matrilineal relatedness. The circumstantial evidence supporting these assessments is persuasive, but confirmation awaits the critical laboratory experiments which are yet to be done.

Studies in the field that indicate a closer link between animal and human language are becoming increasingly common, and few would now support the purely affective interpretation of animal communication. The heretofore unsuspected complexity in the vocal signals that animals use in communicating with conspecifics is being unraveled by controlled observation in natural and quasi-natural habitats. These observations are greatly enhanced by carefully applied technology and the judicious use of call playback experiments (for a review see Cheney and Seyfarth 1990). A striking example is the class of alarm calls used by vervet monkeys that are apparently differentiated on the basis of the type of predator to which they refer (Seyfarth, Cheney, and Marler 1980a,b). It seems likely, on the basis of field evidence, that these animals inform their kin not only of the existence of a predator but of its nature as well. Snakes, leopards, and eagles elicit acoustically different alarm calls from these monkeys, and the evasive action taken by the receivers is adaptive, given the type of predator. Thus, leopard alarm calls lead to running up the nearest tree, while in response to eagle alarm calls, the animals descend from the tree and enter the closest bush. Recorded playback of these calls permitted the observers to at least partially separate the signal from its usual context—that is, the appearance of a predator. In fact, most animals reacted similarly to the recorded calls and took the appropriate evasive action. Cheney and Seyfarth (1990) noted that these different calls were easily distinguished by human listeners and this suggests that the monkeys could also discriminate among them, although the particular experiment has not yet been carried out.

In spite of Lloyd Morgan's canon (Morgan 1894), an entire field of cognitive ethology (see Ristau 1991) has arisen from the ashes of nineteenth century comparative psychology (Romanes 1882, 1883) and is due in no small part to the efforts of the ethology group at Rockefeller University, particularly Griffin (1981). The members of this group have rightfully called attention to the inadequate explanations of the complex behavior observed in other animals in their natural habitats (see Ristau 1991). An integration of cognitive and ethological approaches is an important direction for the study of animal behavior (see Yoerg and Kamil 1991) and, when applied to the comparative method, may improve our understanding of the evolution of behavior and of such complex processes as communication and human language.

Griffin (1991) has argued for the importance of the study of communication, and especially of communicative signals, as objective indicators of the thought processes of other animals. Yoerg and Kamil (1991) take a somewhat more conservative position in an attempt to avoid dealing with mentalistic events such as thinking and awareness and argue for an approach based upon human experimental cognitive psychology. In departing from Griffin's emphasis on communication, they raise the question, "Do words necessarily speak louder than actions?" (Yoerg and Kamil 1991, p. 275). Yoerg and Kamil thus leave the door open for a somewhat more behavioral approach, hopefully not one of those "vestiges of behaviorism that inhibit inquiry" (Griffin 1991, p. 4) and seem to serve as a straw man in much of Griffin's recent writing. But perception is an essential part of the communication process. It is certainly an important element in the study of cognition and of ethology, yet the cognitive ethologists have largely ignored much of the extensive laboratory research on animal perception, which is one of the few areas where quantitative experimental findings could lend credence to their claims of cognitive processing in nonhuman animals (see Berkley and Stebbins 1990; Stebbins and Berkley 1990).

3.3 Laboratory Studies with Species Sounds

Whether human language is an example of saltatory or gradual evolution, its origins, as Darwin (1981 [1871]) argued, lie in the communication systems of other animals. What is more, the findings of Seyfarth, Cheney and Marler, (1980a,b) and Gouzoules, Gouzoules and Marler (1984, 1985) provide persuasive, though not conclusive, evidence that the cognitive structure of human language has its provenance in the referential acoustic signaling, at least of nonhuman primates. That there is still doubt and controversy even among the ethologists themselves is apparent (see Ristau 1991). Communication is a two-way process and it is the signaling side that has received major attention from the ethologists, although the recent playback experiments have begun to focus more on the actions of the receiver. It is in understanding the perception of these complex species signals that laboratory experiments can make a significant contribution. Although playback experiments in the field are able to look at the animal's response to a signal with some of the context removed, in the laboratory investigators can go further and remove the context completely and focus on the perception of the vocal signal itself and the implications for its processing by the auditory system.

One important objective in taking these questions to the laboratory is to determine whether a nonhuman primate's perception of its calls is structured similarly to the manner in which humans perceive speech. Is there evidence that other animals classify and categorize species sounds in the manner that humans categorize speech sounds? For example, human perception of speech shows a remarkable constancy in the face of considerable variance in the acoustic signal. Of course, if pressed, humans can readily tell these sounds apart. Furthermore, for most human listeners, speech seems to be transmitted to and processed in the left hemisphere of the brain. Finally, there are distinctive features in human speech that represent significant information-bearing elements, such as the rapid frequency modulation at the onset of the stop consonants of human speech elements (phonemes), e.g., /ba/ and /da/. The question was whether other primates would demonstrate these perceptual characteristics in the process of perceiving their own species calls. Such an outcome would further strengthen the links between human language, in all of its cognitive complexity, and animal communication.

A somewhat different although related question is whether nonhuman primates perceive human speech sounds as humans do. Some have argued (e.g., Liberman 1982) that "speech is special" and requires unique processors in the nervous system distinct from those that are engaged to deal with nonspeech. If it can be shown in the laboratory that animals discriminate between speech sounds as humans do and categorize them in a like manner, then the argument for the special character of speech is weakened and with it the qualitative distinction between human speech and animal communication. This argument is in no way meant to minimize the considerable differences between speech and animal communication but rather to advance the evolutionary argument that the provenance of language resides in the communication systems of other animals. Laboratory studies of the nonhuman primate's perception of both human speech sounds and species sounds have contributed to the understanding of the processing of biologically relevant complex signals in other animals, of the evolution of communication systems and human language, and of those acoustic features that might provide the cues that enable an animal's identification and discrimination of these stimuli.

With regard to species calls, the study of the perception of complex acoustic signals has greatly benefited from the field studies described above, particularly the early observations of Green (1975b). Green's suggestion that an intimate relationship exists between the acoustic structure of a call and its social function implied that the animals can discriminate among these calls and yet, within certain call classes, perceive these calls similarly. The coo calls, with their relatively simple acoustic structure, offered the opportunity for laboratory experimentation. A continuum was provided by the "smooth early high" (SEH) and "smooth late high" (SLH) coo or contact calls, which differed in the temporal location of the frequency perturbation in the call— early or late. Japanese monkeys (*Macaca fuscata*) and nonspecies controls (other macaques and one nonmacaque species) were first trained in a procedure almost identical to that described in Section 2.2 for frequency discrimination. One signal (either SEH or SLH) is presented monaurally after the subject contacts the response device. After several repetitions (these vary in number from one trial to the next), the other signal is presented and the animal's response is noted. If the animal breaks contact with the response device, it is reinforced with food; if it breaks contact too soon, there is a brief time-out before the next trial begins (Beecher et al. 1979). Early in training, one of each type of signal is presented. When the subject correctly discriminates on most of the trials, another pair of contrasting signals is added, up to a maximum of eight SEHs and seven SLHs. Subjects are required to master the discrimination to a criterion of 90% correct at each stage of the training.

The Japanese monkeys acquired the discrimination quickly at each stage in classifying the calls according to the scheme that Green (1975b) proposed and, in so doing, unequivocally demonstrated perceptual constancy—that is, calls were responded to as SEHs or SLHs in spite of considerable variations, within each category, in the structure of the acoustic signal itself (Beecher et al. 1979; Zoloth et al. 1979). In contrast, the other monkey species had great difficulty at each stage and only learned the discrimination after extensive training. Some of the results are shown in Figure 4.9, which displays the number of trials to mastery at each stage of the discrimination. At this time, only one of the subjects in the control group had learned the discrimination; the others acquired it only after the discrimination had been modified and after many more trials. As a further control, a new group of Japanese and non-Japanese monkeys was trained on a different discrimination. The signals were the same, but the rules and the contingencies of reinforcement were changed. The calls were arbitrarily grouped according to starting frequency —high or low (above or below 600 Hz). In this experiment, the Japanese monkeys had considerable difficulty with the discrimination while the other

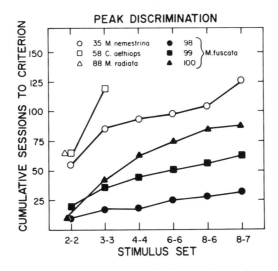

FIGURE 4.9. Cumulative sessions to criterion for 6 animals as a function of the size of the stimulus set. "8-7" (for example) means that there were 8 SLHs and 7 SEHs in the set, "4-4" that there were 4 of each. Thus, the horizontal dimension represents increasing complexity or difficulty of the discrimination. Subject 58 was unable to reach criterion at 3-3, Subject 88 at 2-2; they were later switched to another, easier version of the discrimination (not shown) which they eventually learned. (From Beecher et al. 1979.)

animals found it relatively easy. These experiments in the laboratory confirmed Green's (1975b) hypotheses that the animals were capable of discriminating between the two call types, that they did group these calls into two distinct classes, and that somehow their ability to perceive the calls in this way was a function of their species identity.

As an integral part of these experiments, the stimuli were always given monaurally to the subjects, and the responses were noted according to the ear to which the stimuli were delivered. Should the performance of the Japanese monkeys be superior for signals delivered to the right ear, this would imply a left hemisphere advantage and left hemisphere processing of these species signals. The performance of the Japanese monkeys and one of the controls was significantly better for signals to the right ear; there was no ear advantage for the rest of the controls. In a subsequent series of experiments in a different laboratory and with different subjects, Heffner and Heffner (1984, 1986, 1990) found that left hemisphere lesions in the superior temporal gyrus of Japanese monkeys produced a significant yet temporary decrement in the discrimination between the two call types (SEH and SLH). Right hemisphere lesions had no effect while lesions to both hemispheres produced a permanent loss of the ability to make the discrimination (Heffner and Heffner 1984, 1986, 1990). The results of these two different experiments thus strongly support the hypothesis of cerebral laterality for species-typical signals in a nonhuman primate.

It is argued here that these macaque monkeys, in demonstrating laterality in an area of the brain that in humans processes speech, perceptual constancy, and the ability to categorize meaningful communicative signals from conspecifics, show clear evidence of cognitive strategies in their communication system. This is an example of how monkeys hear their world, and it is evident that these animals use at least some of the same strategies that humans employ in processing speech. The Heffners' findings make it abundantly clear that the neural mechanisms involved in the perception of species calls by Japanese monkeys bear a close anatomical relationship to those found in humans in the perception of speech. It is no longer tenable to support the earlier conceptions of animal calls as little more than reflexive reactions on the part of the receiver to a signaler's emotional or motivational state, which is the classical ethological position that Griffin (1985) had labeled the "groan of pain" interpretation.

On the basis of his observations in the field, Green (1975b) suggested that the temporal location of the frequency perturbation or peak in the coo call enabled the monkeys to discriminate one variant of the call (SEH) from another (SLH). The laboratory studies first confirmed that the animals could discriminate between these variants and, second, supported Green's hypothesis that the position of the frequency peak in the call was its significant enabling feature. It was important to know which acoustic feature(s) in the calls was salient for the discrimination, but another kind of experiment was necessary in order to answer that question. To identify such features, it was necessary to degrade a call by removing certain features while leaving others intact. If, under those conditions, the discrimination broke down, then the feature or features that had been excised would be considered critical. The only practical means of accomplishing this objective would be to computer synthesize the call to make alteration of acoustic features plausible. Before this experiment could be attempted, it was necessary to establish that the monkeys would treat a synthetic call as part of their normal repertoire—that is, like its natural equivalent.

When the monkeys were discriminating between multiple examples of the two call types (SEH and SLH), the probability of reinforcement for correct responses was reduced from 100 to 85% (May, Moody, and Stebbins 1988). This reduction permitted the occasional introduction of new signals without reinforcement so that there was no direction to the subject as to how to categorize the signal. For example, two natural calls (one SEH and one SLH) were presented on several occasions, and the results are seen in Figure 4.10. The SEH call, whose sonogram is on the left, is identified invariably as a SEH while its counterpart (SLH) is almost never identified as a SEH. In Figure 4.11 are shown the results for two synthesized calls, one configured to resemble a SEH, the other a SLH. The animals bought the deception and accepted

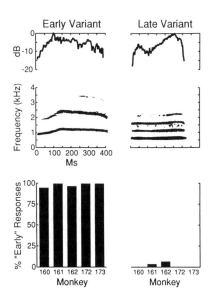

FIGURE 4.10. Sonograms for natural smooth early high (upper left) and smooth late high (upper right) inflected coo calls. The contours at the top of the panel represent the amplitude changes while those just below represent the frequency changes for the fundamental and harmonics. Percent "early" responses to the early inflected call (lower left) and percent "early" responses to the late inflected call (lower right). All animals responded to the smooth early high variant as an early inflected call but not similarly to the late variant. (From May, Moody, and Stebbins 1988.)

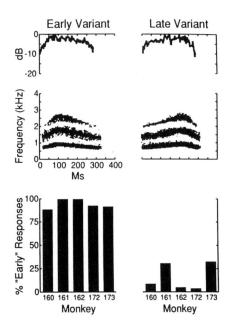

FIGURE 4.11. Sonograms for synthetic early (upper left) and late (upper right) inflected coo calls. The contours at the top of the panel represent the amplitude changes while those just below represent the frequency changes for the fundamental and harmonics. Responses to the calls in lower panels as described for Figure 4.10. Animals responded to the early variant as an early inflected call but not similarly to the late variant. (From May, Moody, and Stebbins 1988.)

the calls as their own, thus permitting the continuation of the experiment and the degradation of the synthetic calls. As noted in the figure, the synthetic SEH was responded to as such almost as frequently as its natural equivalent; the synthetic SLH, on the other hand, was rarely confused with the synthetic SEH.

Since it was suspected that the frequency peak was a critical feature in the call's recognition, the peak was removed from the call but a differential pattern of amplitude modulation early in one call and late in another was retained. Any consistent amplitude cue was rarely seen in either of the two classes of natural calls, although it is certainly an important sign of affect in any animal communication system. The two newly synthesized calls are shown in Figure 4.12, with the animals showing little evidence of discrimination between them. Believing the frequency peak to be important, the calls were altered so as to exaggerate the peak and its location slightly, as shown in Figure 4.13. Four of five animals did not distinguish between the two calls, as shown in the figure. But when the frequency peak was removed and nothing more than a frequency glide, upward for one call and downward for the other, was retained, as shown in Figure 4.14, all subjects discriminated al-

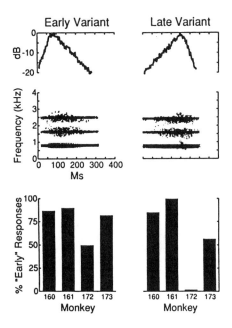

FIGURE 4.12. Sonograms for synthetic early (upper left) and late (upper right) inflected coo calls. The contours at the top of the panel represent the amplitude changes while those just below represent the frequency changes for the fundamental and harmonics. The frequency inflection has been eliminated although the amplitude change has been retained. Responses to the calls in lower panels as described for Figure 4.10. For the most part, the animals responded similarly to both early and late variants. (From May, Moody, and Stebbins 1988.)

most as perfectly as they had done for the natural calls. The conclusion was that the frequency modulation alone was the critical cue for classifying these calls, and the frequency pertubation or reversal of the direction of the modulation, as seen in the peak, was of little significance. This finding is not surprising in view of the importance of frequency modulation in many animal calls and in human speech (see Stebbins and Sommers 1992). We have previously argued for its adaptive value irrespective of the specialized requirements of particular habitats and for its use in increasing the effective signal-to-noise ratio, thus lowering detection thresholds in noisy environments (Stebbins and Sommers 1992). These among other reasons (ibid. for a review of these) substantiate its considerable importance for the perception of communication signals.

Classification or categorization is an efficacious and adaptive way of dealing with many and varied communication signals, especially in animals with relatively complex communication systems. Acoustic signals carrying the same message vary in structure depending on the characteristics of the signaler, signaler-to-receiver distance, social and physical context, and so on. A

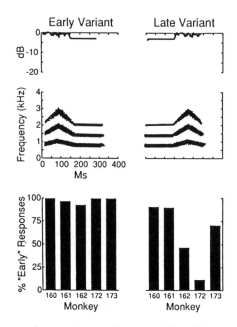

FIGURE 4.13. Sonograms for synthetic early (upper left) and late (upper right) inflected coo calls. The contours at the top of the panel represent the amplitude changes while those just below represent the frequency changes for the fundamental and harmonics. The frequency inflection has been shortened. Responses to the calls in lower panels as described for Figure 4.10. For the most part, the animals responded similarly to both early and late variants. (From May, Moody, and Stebbins 1988.)

particular class of contact or alarm calls should be able to be distinguished from another class of such calls in spite of considerable variation in the properties of the acoustic signal. The message should be clear regardless of who the signaler is, at what distance the signaler might be, or what the current conditions are with regard to terrain, weather, prey, or predator. This is all too evident in human language, and we expect to see its beginnings in the communication systems at least of other primates and perhaps of birds. The ability to respond similarly to signals carrying the same message but varying in their acoustic properties is known as perceptual constancy, and it has been shown that Japanese monkeys demonstrate perceptual constancy in responding similarly to the many varieties of either SEH or SLH contact calls. If these two calls lie on a continuum represented by the temporal location of the frequency peak, then categorization can be sought for based on the animal's partitioning of the continuum such that calls on different sides of the partition are responded to differently while calls on the same side are responded to in the same way—that is, generalization within a class or category and discrimination between classes.

For such an experiment, it is preferable to use synthetic calls because better

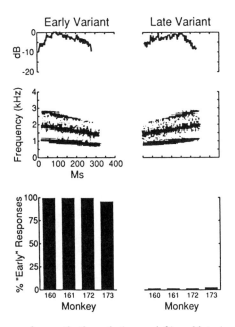

FIGURE 4.14. Sonograms for synthetic early (upper left) and late (upper right) inflected coo calls. The contours at the top of the panel represent the amplitude changes while those just below represent the frequency changes for the fundamental and harmonics. The inflection has been removed from the frequency change. Responses to the calls in lower panels as described for Figure 4.10. Animals responded to the early variant as an early inflected call but not similarly to the late variant. (From May, Moody, and Stebbins 1988.)

control is achieved over the parameters of the call, and then frequency peak location in the call may be varied in isolation. This is a practice modeled after that used in studies of human speech perception. It was possible to use the synthetic calls since the animals treated them in the same way as the natural calls (see above). Two calls near each end of the continuum were used as SEH and SLH. In this way, the discrimination was maintained by reinforcing correct responses to the stimuli. A series of calls was constructed that graded between these two extremes in equal temporal intervals for peak location. Occasionally these calls were individually presented as probes without reinforcement and the subject's response was noted. The results are shown in Figure 4.15. The evidence for categorization of these two calls is clear and robust, with the category partition or boundary placed about halfway along the continuum. While most of the calls are placed unequivocally in one category or the other, there is some ambiguity near the boundary itself at the peak location of 125 ms (May, Moody, and Stebbins 1989).

The series of experiments with the Japanese monkeys has demonstrated several properties of these animals' perception of their own communication

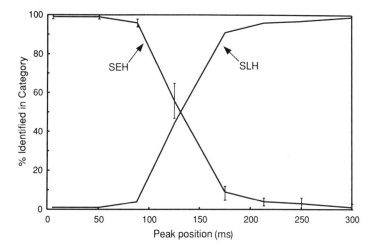

FIGURE 4.15. Identification of synthetic "coo" calls. Data are presented as the mean and the standard error of the mean for four subjects (*Macaca fuscata*). A category boundary, indicated by the sharp transition in the behavioral response, was observed near the midpoint of the function. Although most of the synthetic stimuli were placed in only one of the two vocal categories (smooth early high or smooth late high), responses to the stimulus with a 125-ms peak position were more equivocal. (From May, Moody, and Stebbins 1989.)

sounds that are also evident in our perception of speech: perceptual constancy, categorical perception, neural lateralization, and the importance of frequency modulation as a discriminating feature of the sounds themselves. These common properties argue convincingly for the provenance of human speech and language in the communication signals of other primates. These studies and this view of the origin of language also find support in the results from a distinct set of experiments on the discrimination of human speech and speechlike sounds by nonhuman primates (Hienz and Brady 1988; Kojima et al. 1989; Kuhl and Padden 1982, 1983; Pohl, 1983; Sinnott et al. 1976).

3.4 Laboratory Studies with Human Speech Sounds

Experiments that have investigated the perception of human speech sounds by nonhuman primates have tended to emphasize those aspects of perception that, at one time, were attributed to a speech-specific mode of processing in humans. The hypothesis underlying these studies is that if animals who do not possess language nonetheless demonstrate perceptual processing of speech stimuli similar to that employed by humans, then it is not reasonable to attribute such processing to linguistic specialization, but rather to assume that it is based on some more general characteristic of the mammalian auditory system. In large part, such studies, which have dealt primarily with categorical perception, have demonstrated considerable cross-species similarity in the perception of human speech and, as a result, have led to a rethinking of the nature and existence of a uniquely human speech processing mode (see, for example, Morse and Snowdon 1975; Sinnott et al. 1976; Waters and Wilson 1976; Kuhl and Padden 1982, 1983). The rapid categorization of signals with similar function but different acoustic structure is an effective adaptive strategy since it readily facilitates the communication process. These studies have resulted in a reevaluation of the manner in which nonhuman communication systems might be organized and of the potential cues that are available to animals for use in acoustic signaling.

The strict definition of categorical speech perception requires two features: first, that stimuli be perceived as being members of discrete categories, even though they are acoustically different; and second, that discrimination acuity be greater for stimuli having a given acoustic difference but drawn from different categories than stimuli separated by the same acoustic difference but drawn from the same category. One of the best examples of a study that demonstrates both of these conditions is by Kuhl and Padden (1983) in which macaque monkeys (*Macaca fuscata* and *Macaca nemestrina*) were trained on a same/different task and then tested with a variety of stimulus pairs drawn from the /bae-dae-gae/ continuum which is produced by differences in the place of articulation of the initial stop consonant. The results of that study showed labeling functions with abrupt transitions between the various phonetic categories and discrimination functions with enhanced sensitivity to stimulus differences at the location of the transition points, thus fulfilling both requirements for a demonstration of categorical perception.

Several other studies have shown categorical-like labeling functions but, in other ways, presented results that question whether the animal subjects were truly demonstrating humanlike categorical perception. Sinnott et al. (1976), for example, showed that monkeys (*Macaca fuscata*, *Cercopithecus aethiops*, and *Macaca fascicularis*) produced a category boundary on the /ba-da/ continuum that was close to that produced by humans. However, when reaction times to stimulus differences were examined in this experiment, the monkey subjects showed graded responses along the continuum while humans showed no such gradation, as would be predicted if they were unable to discriminate within the categories. Waters and Wilson (1976) also demonstrated category boundaries for *Macaca mulatta* near those shown by humans on the /ba-pa/ voice-onset-time continuum and, in addition, showed improved discrimination between categories relative to that shown within categories. Although these results appear to meet the strict definition of categorical perception, they also illustrate a pitfall of perceptual research with animals in that they demonstrate a shift in the category boundary when the range of stimuli used in training was varied. Such a shift is not shown by humans, and its presence would suggest that the categorization may be the result of the conditions of training rather than representing a "natural" categorization.

Another means of demonstrating categorization of speech sounds is through the use of the statistical techniques of multidimensional scaling (MDS) and cluster analysis. These techniques rely on a behavioral measure of similarity between pairs of stimuli; usually response latency (reaction time) to stimulus change is used. The greater the perceived difference, the faster the subject will respond, while small differences result in long latencies. One of the major advantages of these procedures is that any response made to a stimulus difference can be reinforced, and therefore, there is no danger that the observed categorization will be the result of the experimenter-defined conditions of training. Since between-category differences should be perceived as being greater than those within categories, it is reasonable to expect that these techniques would correctly sort categorically perceived stimuli. The MDS and cluster analysis procedures were applied in a study by Kojima et al. (1989) using chimpanzees (*Pan troglodytes*) as subjects. The results of this study demonstrated similar separation of stop consonants into voiced vs. voiceless groupings for humans and apes, but there were considerable individual differences within each of the groupings. For example, the MDS procedure failed to show any separation between /da/ and /ga/ for one of the chimpanzee subjects.

One of the earliest demonstrations that human speech might involve processing mechanisms different from those evoked by simpler stimuli was carried out by Dewson, Pribram, and Lynch (1969). Monkeys (*Macaca mulatta*) were first trained to discriminate between a tone and a noise, the vowel sounds /i/and /u/, or a pair of visual stimuli. When they had mastered each of these discrimination problems, portions of their auditory cortex were surgically removed. When retested, those subjects with lesions of their primary auditory cortex were unable to relearn the vowel-sound discrimination but had no difficulty with either the visual discrimination or the tone-noise problem. Dewson's findings on *Macaca mulatta*, like those of the Heffners (1984, 1986, 1990) on *Macaca fuscata*, emphasize the specificity of the temporal cortex in processing acoustic communication signals and further suggest that it may be specific features of such signals, or of particular contrasts between such signals, that require cortical processing.

Hemispheric specialization for processing human speech sounds has also been studied in nonhuman primates (Pohl 1983), with somewhat confusing results relative to the specializations cited previously (see Section 3.3) for monkeys discriminating conspecific communication signals. In Pohl's study, baboons (*Papio cynocephalus*) were trained to discriminate between pure tones, musical chords, vowel sounds, and consonant-vowel syllables. Although all subjects showed ear advantages for some stimuli, suggesting differential hemispheric processing, most were a left-ear advantage as opposed to the right-ear advantage shown by monkeys discriminating conspecific communication signals. While this difference may not be significant, it is noteworthy that both speech and nonspeech tasks produced similar ear advantage results, and, in fact, for all subjects, the best predictor of ear advantage for the consonant-vowel task was the ear advantage seen with the musical chord discrimination. A possible, although probably premature, interpretation of the hemispheric specialization results taken as a whole is that hemispheric specialization exists in nonhuman primates for processing acoustic stimuli, with communicatively relevant conspecific communication contrasts having privileged access to left hemisphere processing.

One other area of the study of human speech perception that has been extended to nonhuman primates is that of the perception of vowel sounds. The classification of these signals by humans seems to be controlled by the position and amplitude of the peaks in their acoustic frequency spectrum (e.g., Peterson and Barney 1952). Studies in which baboons (*Papio anubis*; Hienz and Brady 1988) and chimpanzees (*Pan troglodytes*; Kojima and Kiritani 1989) were trained to discriminate between a variety of human vowels have shown similar correlations between vowel spectrum and discriminability, although not necessarily the same as that shown by humans. Kojima and Kiritani, for example, showed that the humans discriminating /i/ from /u/ attended to differences in the second spectral peak (formant), while chimpanzees had some difficulty discriminating between these vowels, apparently because they were attending to the first spectral peaks that were nearly identical in the two stimuli.

Recently, Sommers et al. (1991, 1992) have examined the discrimination of vowel sounds by monkeys (*Macaca fuscata*) in some detail in an attempt to isolate the acoustic cues used in making the discriminations. They pointed out that vowel sounds, like any voiced speech, are broadband sounds consisting of multiple harmonics of the voicing frequency, which, for an adult male talker, is in the range of 100 Hz. The spectrum of this broadband sound, that is, the amplitude of each of the harmonics, is determined by the shape of the vocal tract that acts as an acoustic filter. As the configuration of the vocal tract is changed, the center frequencies of the acoustic filters shift, causing some harmonics to increase in amplitude while others decrease. Sommers et al. (1992) examined the ability of monkeys to detect small changes in the center frequency of the vocal tract filters by training subjects to discriminate between computer-synthesized vowel sounds that differed in the frequency of either the first or second formant. They demonstrated that monkeys showed greater acuity for a shift in the center frequency of a particular formant than for shifts in the frequency of a pure tone at the same frequency. Further analysis revealed that the animals were basing their discriminations on the changes in amplitude of the component harmonics that occurred as the center frequency of the formant was shifted and, further, that a small change in only one or two of these harmonics was sufficient to explain the acuity shown by the subjects. These results pointed out that, although experimenters may specify complex stimuli as changing along one dimension such as frequency, subjects may, in fact, be using correlated cues from a different dimension, in this case, the relative intensity of component harmonics. Furthermore, the ability to utilize these cues may reflect particularly well-developed acuity for the controlling dimension, suggesting that it is highly salient in the communication systems of the subject species.

In Section 3, we presented evidence that the communication signals of nonhuman primates contain a richness of cognitive content that had gone without notice until it was revealed by detailed studies both in the laboratory and in the field. We used this evidence to argue that the uniqueness that had previously been attributed to human speech perception was not unique at all but, rather, was mirrored in many of the perceptual aspects of the acoustic communication systems of nonhumans as well. In this section, we have extended those arguments to show that many of the perceptual specializations for dealing with human speech signals themselves are also mirrored in nonhumans. These findings add to our argument that human speech and language may have their evolutionary roots in the communication signals of other species.

4. Summary

We have reviewed much of what is known about auditory perception in those species of nonhuman primates that have been tested in the laboratory. We have examined auditory acuity for simple signals (pure tones) in an attempt to define the outer limits of the auditory perceptual space or the acoustic dimensions within which these animals operate. The dimensions are very similar to those that characterize human hearing, with at least two significant exceptions. The hearing range of nonhuman primates extends to between 30 and 60 kHz, which places them between humans (with an upper limit at about 20 kHz) and most other mammals (at 45 to 90 kHz). The discriminative acuity of nonhuman primates, particularly for frequency, reveals somewhat less sensitivity as compared to humans. Whether, as seems likely, this difference reflects the more intense selective pressures on human primates that are related to a much more complex form of intraspecific acoustic communication (spoken language), it is not possible to say. However, it is important to note that the similarities between human and nonhuman primates with regard to what they can hear far outweigh the differences, which argues convincingly for common physiological mechanisms.

Unfortunately, little is known about the perception of complex signals such as those used in intraspecific communication in nonhuman primates. This significant area of laboratory experimentation remains largely untapped, although recent field studies with well-controlled playback experiments provide fertile ground for further laboratory research. Based on results from one species (*Macaca fuscata*), evidence was found that these animals perceive communication signals, either those uttered by a conspecific or those common to human speech, much as humans perceive their own speech. Perceptual constancy, categorical perception, hemispheric laterality, and attention to common acoustic features not only bespeak the origins of human language in the calls of other animals but provide evidence for cognition and its neural basis in these animals, qualities or properties heretofore reserved only for their human counterparts. We can only predict that further studies will tighten this linkage between human and nonhuman primate communication and suggest ways in which we can better understand the evolutionary process that led to language and the cognitive processing of biologically relevant signals in both humans and nonhumans.

References

Arlinger SD, Jerlvall LB, Ahren T, Holmgren EC (1977) Thresholds for linear frequency ramps of a continuous pure tone. Acta Otolaryngol 83:317–327.

Beecher MD (1974a) Hearing in the owl monkey (*Aotus trivirgatus*): Auditory sensitivity. J Comp Physiol Psychol 86:898–901.

Beecher MD (1974b) Pure tone thresholds of the squirrel monkey (*Saimiri sciureus*). J Acoust Soc Am 55:196–198.

Beecher MD, Petersen MR, Zoloth SR, Moody DB, Stebbins WC (1979) Perception of conspecific vocalizations by Japanese macaques: Evidence for selective attention and neural lateralization. Brain Behav Evol 16:443–460. Karger, Basel.

Berkley MA, Stebbins WC (1990) Comparative Perception, Volume 1. New York: John Wiley and Sons.

Brown CH (1976) Auditory localization in primates: The role of stimulus bandwidth. Doctoral Dissertation, Michigan State University, East Lansing, MI.

Brown CH, May BJ (1990) Sound localization and binaural processes. In: Berkley MA, Stebbins WC (eds) Comparative Perception, Volume 1. New York: John Wiley and Sons, pp. 247–284.

Brown CH, Waser PM (1984) Hearing and communication in blue monkeys (*Cercopithecus mitis*). Anim Behav 32:66–75.

Brown CH, Beecher MD, Moody DB, Stebbins WC (1980) Localization of noise bands by Old World monkeys. J Acoust Soc Am 68:127–132.

Cheney DL, Seyfarth RM (1990) How Monkeys See the World: Inside the Mind of Another Species. Chicago: University of Chicago Press.

Clark WW, Clark CL, Moody DB, Stebbins WC (1974) Noise-induced hearing loss in the chinchilla determined by a positive reinforcement technique. J Acoust Soc Am 56:1202–1209.

Darwin C (1981) The Descent of Man and Selection in Relation to Sex. Princeton, NJ: Princeton University Press. (Photoreproduction of 1871 edition published by J. Murray, London.)

Dewson JH III, Pribram KH, Lynch JC (1969) Effects of ablations and temporal cortex upon speech sound discrimination in monkey. Exp Neurol 24:579–591.

Donisthorpe JH (1958) A pilot study of the mountain gorilla (*Gorilla gorilla beringei*) in South West Uganda, February to September 1957. Afr J Sci 54:195–217.

Elder JH (1933) Audiometric studies with the chimpanzee. Psychol Bull 30:547–548.

Elder JH (1934) Auditory acuity of the chimpanzee. J Comp Psychol 17:157–183.

Elder JH (1935) The upper limit of hearing in chimpanzee. Am J Physiol 112:109–115.

Farrer DN, Prim MM (1965) A preliminary report on auditory frequency threshold comparisons of humans and preadolescent chimpanzees. Technical Report No. 65–6, 6571st Aeromedical Research Laboratory, Holloman Air Force Base, New Mexico.

Fay RR (1988) Hearing in Vertebrates: A Psychophysics Databook. Winnetka, ILL: Hill-Fay Associates.

Filling S (1958) Difference Limen for Frequency. Andelsbogtrykkeriet, 1 Odense, Denmark.

Gouzoules H, Gouzoules S (1989) Design features and developmental modification of pigtail macaque (*Macaca nemestrina*) agonistic screams. Anim Behav 37:383–401.

Gouzoules H, Gouzoules S, Marler P (1985) External reference and affective signaling in mammalian vocal communication. In: Zivin G (ed) The Development of Expressive Behavior, Biology-Environment Interactions. Orlando, FL: Academic Press, pp. 77–101.

Gouzoules S, Gouzoules H, Marler P (1984) Rhesus monkey (*Macaca mulatta*) screams: Representational signalling in the recruitment of agonistic aid. Anim Behav 32:182–193.

Green S (1975a) Auditory sensitivity and equal loudness in the squirrel monkey (*Saimiri sciureus*). J Exp Anal Behav 23:255–264.

Green S (1975b) Variation of vocal pattern with social situation in the Japanese monkey (*Macaca fuscata*): A field study. In: Rosenblum L (ed) Primate Behavior, Volume 4. New York: Academic Press, pp. 1–102.

Griffin DR (1981) The Question of Animal Awareness. New York: Rockefeller. University Press.

GriffinDR (1985) Animal consciousness. Neurosci Biobehav Rev 9:615–622.

Griffin DR (1991) Progress toward a cognitive ethology. In: Ristau CA (ed) Cognitive Ethology: The Minds of Other Animals. Hillsdale, NJ: Lawrence Erlbaum Associates, pp. 3–17.

Harrison JM, Downey P, Segal M, Howe M (1971) Control of responding by the location of auditory stimuli: Rapid acquisition in monkeys and rats. J Exp Anal Behav 15:379–386.

Heffner HE, Heffner RS (1984) Temporal lobe lesions and perception of species-specific vocalizations by macaques. Science 226:75–76.

Heffner HE, Heffner RS (1986) Effect of unilateral and bilateral auditory cortex lesions on the discrimination of vocalizations by Japanese macaques. J Neurophysiol 56:683–701.

Heffner HE, Heffner RS (1990) Role of primate auditory cortex in hearing. In: Stebbins WC, Berkley MA (eds) Comparative Perception, Volume 2. New York: John Wiley and Sons, pp. 279–310.

Heffner HE, Masterton RB (1970) Hearing in primitive primates: Slow loris (*Nycticebus coucang*) and potto (*Perodicticus potto*). J Comp Physiol Psychol 71:175–182.

Heffner HE, Ravizza R, Masterton RB (1969a) Hearing in primitive mammals. III: Tree shrew (*Tupaia glis*). J Aud Res 9:12–18.

Heffner HE, Ravizza R, Masterton RB (1969b) Hearing in primitive mammals. IV: Bush baby (*Galago senegalensis*). J Aud Res 9:19–23.

Heffner RS, Masterton RB (1990) Sound localization in mammals: Brainstem mechanisms. In: Berkley MA, Stebbins WC (eds) Comparative Perception, Volume 1. New York: John Wiley and Sons, pp. 285–314.

Hienz RD, Brady JV (1988) The acquisition of vowel discriminations by nonhuman primates. J Acoust Soc Am 84:186–194.

Jesteadt W, Wier CC, Green DM (1977) Intensity discrimination as a function of frequency and sensation level. J Acoust Soc Am 61:169–177.

Kojima S (1990) Comparison of auditory functions in the chimpanzee and human. Folia Primatol 55:62–72.

Kojima S, Kiritani S (1989) Vocal-auditory functions in the chimpanzee: Vowel perception. Int J Primatol 10:199–213.

Kojima S, Tatsumi IF, Kiritani S, Hirose H (1989) Vocal-auditory functions in the chimpanzee: Consonant perception. Hum Evol 4:403–416.

Kuhl PK, Padden DM (1982) Enhanced discrimination at the phonetic boundaries for the voicing feature in macaques. Percept Psychophys 32:542–550.

Kuhl PK, Padden DM (1983) Enhanced discrimination at the phonetic boundaries for place feature in macaques. J Acoust Soc Am 73:1003–1010.

Liberman AM (1982) On finding that speech is special. Am Psychol 37:148–167.

Marler P (1955) Characteristics of some animal calls. Nature 176:6–8.

Marler P (1957) Specific distinctiveness in the communication signals of birds. Behaviour 11:13–39.

Masterton RB, Heffner HE, Ravizza R (1969) The evolution of human hearing. J Acoust Soc Am 45:966–985.

May BJ, Moody DB, Stebbins WC (1988) The significant features of Japanese macaque coo sounds: A psychophysical study. Anim Behav 36:1432–1444.

May BJ, Moody DB, Stebbins WC (1989) Categorical perception of conspecific communication sounds by Japanese macaques, *Macaca fuscata*. J Acoust Soc Am 85: 837–847.

Moody DB, Stebbins WC (1989) Salience of frequency modulation in primate communication. In: Hulse SH, Dooling RJ (eds) The Comparative Psychology of Audition: Perceiving Complex Sound. Hillsdale, NJ: Lawrence Erlbaum Associates, pp. 353–376.

Moody DB, Beecher MD, Stebbins WC (1976) Behavioral methods in auditory research. In: Smith C, Vernon J (eds) Handbook of Auditory and Vestibular Research Methods. Springfield, IL: Charles C Thomas, pp. 439–497.

Moody DB, May BJ, Cole DM, Stebbins WC (1986) The role of frequency modulation in the perception of complex stimuli by primates. Exp Biol 45:219–232.

Morgan CL (1894) Introduction to Comparative Psychology. New York: Scribners.

Morse PA, Snowdon CT (1975) An investigation of categorical speech discrimination by rhesus monkeys. Percept Psychophys 17:9–16.

Nordmark J (1968) Mechanisms of frequency discrimination. J Acoust Soc Am 44: 1533–1540.

Olsho LW, Koch EG, Carter EA (1988) Nonsensory factors in infant frequency discrimination. Infant Behav Dev 11:205–222.

Peterson GE, Barney HL (1952) Control methods used in the study of vowels. J Acoust Soc Am 24:175–184.

Pfingst BE, Hienz R, Kimm J, Miller J (1975) Reaction-time procedure for measurement of hearing. I. Suprathreshold functions. J Acoust Soc Am 57:421–430.

Pohl P (1983) Central auditory processing. V. Ear advantages for acoustic stimuli in baboons. Brain and Language 20:44–53.

Prosen CA, Moody DB, Sommers MS, Stebbins WC (1990) Frequency discrimination in the monkey. J Acoust Soc Am 88:2152–2158.

Riesz RR (1928) Differential intensity sensitivity of the ear for pure tones. Phys Rev 31:867–875.

Ristau CA (1991) Cognitive Ethology: The Minds of Other Animals. Hillsdale, NJ: Lawrence Erlbaum Associates.

Romanes GJ (1882) Animal Intelligence. London: Kegan Paul, Trench & Co.

Romanes GJ (1883) Mental Evolution in Animals. London: Kegan Paul, Trench & Co.

Schaller GB (1963) The Mountain Gorilla: Ecology and Behavior. Chicago: University of Chicago Press.

Serafin JV, Moody DB, Stebbins WC (1982) Frequency selectivity of the monkey's auditory system: Psychophysical tuning curves. J Acoust Soc Am 71:1513–1518.

Seyfarth RM, Cheney DL, Marler P (1980a) Monkey responses to three different alarm calls: Evidence of predator classification and semantic communication. Science 210:801–803.

Seyfarth RM, Cheney DL, Marler P (1980b) Vervet monkey alarm calls: Semantic communication in a free-ranging primate. Anim Behav 28:1070–1094.

Sinnott JM, Beecher MD, Moody DB, Stebbins WC (1976) Speech sound discrimination by monkeys and humans. J Acoust Soc Am 60:687–695.

Sinnott JM, Petersen MR, Hopp SL (1985) Frequency and intensity discrimination in humans and monkeys. J Acoust Soc Am 78:1977–1985.

Sinnott JM, Owren MJ, Petersen MR (1987) Auditory frequency discrimination in primates: Species differences (Cercopithecus, Macaca, Homo). J Comp Psychol 101: 126–131.

Sivian LJ, White SD (1933) On minimum audible sound fields. J Acoust Soc Am 4: 288–321.

Smith DW, Moody DB, Stebbins WC (1987) Effects of changes in absolute signal level on psychophysical tuning curves in quiet and noise in the patas monkey. J Acoust Soc Am 82:63–68.

Smith DW, Moody DB, Stebbins WC (1990) Auditory frequency selectivity. In: Berkley MA, Stebbins WC (eds) Comparative Perception, Volume 1. New York: John Wiley and Sons, pp. 67–95.

Sommers MS, Moody DB, Prosen CA, Stebbins WC (1991) Spectral shape discrimination by Japanese monkeys. Abstracts of the Fourteenth Midwinter Research Meeting of the Association for Research in Otolaryngology 471.

Sommers MS, Moody DB, Prosen CA, Stebbins WC (1992) Formant frequency discrimination by Japanese macaques (Macaca fuscata). J Acoust Soc Am 91:3499–3510.

Stebbins WC (1966) Auditory reaction time and the derivation of equal loudness curves for the monkey. J Exp Anal Behav 9:135–142.

Stebbins WC (1970) Animal Psychophysics: The Design and Conduct of Sensory Experiments. New York: Appleton-Century-Crofts.

Stebbins WC (1971) Hearing. In: Schrier AM, Stollnitz F (eds) Behavior of Nonhuman Primates, Volume 3. New York: Academic Press, pp. 159–192.

Stebbins WC (1973) Hearing of Old World monkeys (Cercopithecinae). Am J Phys Anthropol 38:357–364.

Stebbins WC (1975) Hearing of the anthropoid primates: A behavioral analysis. In: Eagles EL (ed) The Nervous System: Human Communication and Its Disorders, Volume 3. New York: Raven Press, pp. 113–124.

Stebbins WC (1978) Hearing of the primates. In: Chivers DJ, Herbert J (eds) Recent Advances in Primatology. New York: Academic Press, pp. 703–720.

Stebbins WC, Berkley MA (1990) Comparative Perception, Volume 2. New York: John Wiley and Sons.

Stebbins WC, Sommers MS (1992) Evolution, perception, and the comparative method. In: Popper AN, Webster DB, Fay RR (eds) The Evolutionary Biology of Hearing. New York: Springer-Verlag, pp. 211–227.

Stebbins WC, Green S, Miller FL (1966) Auditory sensitivity of the monkey. Science 153:1646–1647.

Stebbins WC, Brown CH, Petersen MR (1984) Sensory processes in animals. In: Darian-Smith I, Brookhart J, Mountcastle VB (eds) Handbook of Physiology, Sensory Functions, Volume 1. Bethesda, MD: American Physiological Society, pp. 123–148.

Tyler RS, Wood EJ, Fernandes M (1983) Frequency resolution and discrimination of constant and dynamic tones in normal and hearing-impaired listeners. J Acoust Soc Am 74:1190–1199.

Waters RA, Wilson WA Jr (1976) Speech perception by rhesus monkeys. The voicing distinction in synthesized labial and velar stop consonants. Percept Psychophys 19:285–289.

Wendt GR (1934) Auditory acuity of monkeys. Comp Psychol Monogr 10(4).

Wier CC, Jesteadt W, Green DM (1977) Frequency discrimination as a function of frequency and sensation level. J Acoust Soc Am 61:178–184.

Yoerg SI, Kamil AC (1991) Integrating cognitive ethology with cognitive psychology. In: Ristau CA (ed) Cognitive Ethology: The Minds of Other Animals. Hillsdale, NJ: Lawrence Erlbaum Associates, pp. 271–289.

Zoloth SR, Petersen MR, Beecher MD, Green S, Marler P, Moody DB, Stebbins WC (1979) Species-specific perceptual processing of vocal sounds by monkeys. Science 204:870–873.

5
Structure of the Mammalian Cochlea

Stephen M. Echteler, Richard R. Fay, and Arthur N. Popper

1. Introduction

Two features distinguish mammalian hearing from auditory reception in fishes, amphibians, reptiles, and birds. First, the audible frequency range is significantly broader in mammals than in other vertebrate taxa due to the responsiveness of the mammalian ear to higher frequency sounds. Second, when compared with other vertebrates, mammals show considerably more species diversity in particular attributes of hearing such as the lower frequency limit of hearing, the upper frequency range of hearing, and the frequency of maximum sensitivity (Fay 1988). Some mammals, which might be termed hearing generalists, perceive a broad range of sound frequencies and show little variation in threshold sensitivity throughout a large portion of their hearing range. In contrast, other species, which could be called hearing specialists, respond to sounds within a more restricted bandwidth and generally display a greater sensitivity for particular frequencies, often those found within behaviorally relevant acoustic signals.

Identification of the structural correlates that underlie the functional diversity in mammalian hearing requires comparative anatomical studies of the ears in a wide range of mammals. Growing evidence suggests that the frequency range and auditory thresholds exhibited by a particular species may be largely predictable from the transmission and filter characteristics imposed by the structures of its external and middle ears (Plassman and Brändle 1992; Rosowski 1992, Chapter 6). Yet before these externally filtered acoustic signals can be perceived as sound, they must first be transduced into electrical signals by the inner ear and relayed to the auditory central nervous system. Consequently, it is not surprising that considerable differences have also been observed in the cochlear structure within individual animals, between those regions of the inner ear that encode high vs. low frequencies, and among species that are specialized for hearing in particular frequency ranges.

The purpose of this chapter is to provide a general introduction to the cellular anatomy of the mammalian cochlea and to present an overview of the structural variations observed in this cellular bauplan using selected examples of both hearing generalists and specialists. The goal of this comparative approach is to identify modifications of specific cochlear structures which might contribute to differences in cochlear function.

This review is, by necessity, limited in scope. It is not meant to be an exhaustive survey of species differences in cochlear structure. Nor does it cover species differences in cochlear innervation or the details of cochlear ultrastructure. Those interested in these topics are referred to several excellent recent reviews (Kiang et al. 1984; Lim 1986; Ryugo 1992). An additional discussion of the inner ear that is germane to comparative issues is that of Vater (1988) on bats.

During the course of the chapter, we will be referring to a large number of mammalian species. The scientific names of these animals, along with basic data on their cochleae, are given in Table 5.1.

2. Structural Overview of the Mammalian Labyrinth

In this section, we provide a basic overview of the gross structure of the mammalian ear. For the most part, the overall structure of the ear is essentially similar in all mammals, and the general structures described here are applicable to most species. Additional detailed information about the gross structures of the ear can be found in Bloom and Fawcett (1975).

2.1 Bony Labyrinth

The inner ear of mammals, termed the labyrinth because of its intricate architecture, is composed of a connected series of canals and cavities carved from the petrous portion of the temporal bone (stippled region in Fig. 5.1) Centrally located within this complex lies the vestibule, a large chamber that separates the bony labyrinth into a posterior (vestibular) portion comprised of three semicircular canals and an anterior (auditory) portion consisting of a coiled cochlea. This entire structure is filled with perilymph, a filtrate of cerebrospinal fluid that flows into the labyrinth via the perilymphatic duct, a canal extending from the subarachnoid space around the brain into the vestibule. Perilymph, like extracellular fluid, is especially rich in sodium ions ($Na^+ = 140$ mM) but relatively poor in potassium ions ($K^+ = 4$ mM; Johnstone and Sellick 1972).

Within the bony vestibule, and facing the middle ear cavity, are two membranous windows. The larger of these, the oval window, is attached to the footplate of the stapes and transmits the tympanum-induced movements of the middle ear bones into pressure changes within the fluid-filled inner ear. The round window, a smaller flexible membrane located at the base of the bony cochlea near its juncture with the vestibule, moves in opposite relation

TABLE 5.1. Comparative anatomy of the cochlear duct.

Species	Body weight (kg)	Audible frequency range (kHz)	Audible frequency range (Octaves)	Number of cochlear turns	Length of basilar membrane (mm)	Width of basilar membrane (μm)	Thickness of basilar membrane (μm)*	"Stiffness" of basilar membrane (T/W)**	Authors
Elephant (Elephas maximus)	4000	0.18 to 5.7	5.0	2.25	60				West 1985
Mole rat (Spalax ehrenbergi)	0.08	0.1 to 10.0	6.6	3.5	13.7	apex: 200 base: 120	apex: 18.0 base: 9.0	apex: 0.09 base: 0.075	Bruns et al. 1988
Human (Homo sapiens)	75	0.13 to 16.0	7.0	2.75	35	apex: 500 base: 100			Nadol 1988
Common marmoset (Callithrix jacchus)	0.35	0.22 to 18.0	6.4	2.75					West 1985
Horse (Equus caballus)	600	0.2 to 22.0	6.8	2.5					West 1985
Cow (Bos taurus)	500	0.14 to 22.0	7.3	3.5	38				West 1985
Chinchilla (Chinchilla langer)	0.8	0.09 to 25.0	8.1	3.0	18.5	apex: 310 base: 248	apex: 5.50 base: 15.0	apex: 0.018 base: 0.060	Lim 1980
Kangaroo rat (Dipodomys merriami)	0.05	0.1 to 25.0	8.0	3.5	9.83	apex: 254 base: 100	apex: 46.4 base: 8.60	apex: 0.183 base: 0.086	Webster and Webster 1977
Rabbit (Oryctolagus cuniculus)	2	0.45 to 36.0	6.3	2.25	14.5–16				West 1985
Dog (Canis familiaris)	15	0.2 to 36.0	7.5	3.25					West 1985
Raccoon (Procyon lotor)	10	0.14 to 37.0	8.0	2.5					West 1985
Sheep (Ovis aries)	50	0.4 to 37.0	6.5	2.25					West 1985
Gerbil (Meriones unguiculatis)	0.05	0.25 to 45.0	7.4	3.25	12.1	apex: 250 base: 100	apex: 35.0 base: 10.0	apex: 0.140 base: 0.100	Plassmann, Peetz, and Schmidt 1987

Guinea pig (Cavia)	0.5	0.2 to 45.0	7.8	4.0	18.8	apex: 250 base: 70	apex: 1.34 base: 7.40	apex: 0.005 base: 0.106	Fernández 1952
Seal (Phoca vitulina)	50	0.49 to 58.0	6.9	2.25					West 1985
Cat (Felis catus)	2.5	0.125 to 60.0	8.9	3.0	22.5	apex: 370 base: 80			Nadol 1988
Rat (Rattus norvegicus)	0.2	1.0 to 59.0	5.8	2.2	10.7	apex: 189 base: 59	apex: 2.0 base: 17.7	apex: 0.01 base: 0.30	Burda, Ballast, and Bruns 1988
Mouse (Mus musculus)	0.01	5.0 to 60.0	3.5	2.0	6.8	apex: 160 base: 40	apex: 1.0 base: 14.5	apex: 0.013 base: 0.363	Ehret and Frankenreiter 1977
Horseshoe bat (Rhinolophus ferrumequinum)	0.02	7.0 to 90.0	3.6	3.5	16.1	apex: 150 base: 80	apex: 2.0 base: 34.0	apex: 0.013 base: 0.425	Bruns 1976
Little brown bat (Myotis lucifugus)	0.007	12.5 to 100	3.0	2.25					West 1985
Dolphin (Tursiops truncatus)	120	0.4 to 145	8.4	2.0	38.5	apex: 350 base: 25	apex: 5.0 base: 25.0	apex: 0.014 base: 1.0	Wever et al. 1971b

* Measurements of pars pectinata component of basilar membrane
** T = thickness of pars pectinata component of basilar membrane
W = width of entire basilar membrane (pars arcuta and pars pectinata)

to the displacements of the oval window and serves as a pressure release mechanism for the incompressible fluid within the bony labyrinth (Wever and Lawrence 1954; Békésy 1960).

2.2 Membranous Labyrinth

Encased within the bony labyrinth, and conforming to its general shape, lies a connected system of ducts and pouches fashioned from epithelial tissue (Fig. 5.1). This membranous labyrinth is attached to the overlying bone through trabeculae of connective tissue and is filled with endolymph. Unlike perilymph, endolymph resembles intracellular fluid in its composition, being especially rich in potassium ions ($K^+ = 120$ mM) and poor in sodium ions ($Na^+ = 1$ mM; Bosher and Warren 1968). Endolymph is produced locally within the membranous labyrinth through the action of specialized dark cells within the vestibular epithelium and by the stria vascularis of the cochlear duct. It is drained from the membranous labyrinth via the endolymphatic

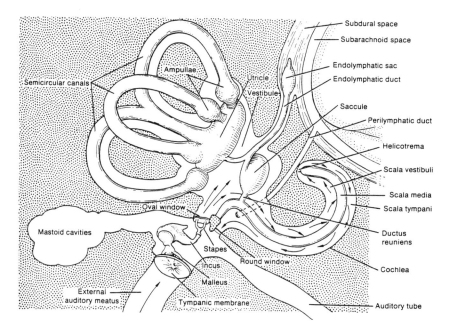

FIGURE 5.1. Gross anatomy of the mammalian inner ear. Depicted in detail is the membranous labyrinth, a connected system of ducts and pouches that contains the sensory endorgans mediating vestibular and auditory functions. Surrounding the membranous labyrinth and outlining its general shape is the bony labyrinth, a series of canals and cavities carved from the temporal bone of the skull (stippled region). The arrows indicate the direction of fluid flow within the bony cochlea following an inward deflection of the oval window. (From Bast and Anson. The Temporal Bone and the Ear 1949, Courtesy of Charles C Thomas, Publisher, Springfield, Illinois).

duct and resorbed by the endolymphatic sac, a blind pouch which terminates within the subdural space surrounding the brain.

Vestibular and auditory receptors are spatially segregated within the membranous labyrinth and are located within thickened regions of specialized epithelium. Two types of vestibular endorgans are found within the posterior labyrinth. The first type, the cristae, are found at the base of each semicircular canal where the canal joins the vestibule. The cristae are positioned within expansions of the membranous labyrinth, called ampullae, which join the semicircular canal ducts to the utriculus. The cristae detect fluid motions within the semicircular canals that are induced by angular accelerations of the head (Lowenstein and Sand 1940). Another type of vestibular endorgan, the maculae, are found within the saccule and utricle, cavities of the membranous labyrinth that are joined together by the endolymphatic duct and housed within the vestibule. The maculae detect positions, tilts, and linear accelerations of the head (Lowenstein and Roberts 1949). A detailed description of the structural and functional features of these vestibular sensory endorgans is beyond the scope of this chapter but can be found in several excellent reviews (Smith and Tanaka 1975; Goldberg and Fernández 1984).

The anterior (auditory) portion of the membranous labyrinth consists of the cochlear duct, which originates within the vestibule between the oval and round windows and is joined to the saccule by the ductus reuniens. Extending forward from the vestibule, the cochlear duct winds, like a mountain road, around the modiolus, a cone of spongy bone. Both the length of the cochlear duct and the number of turns it displays vary widely among species. When viewed in cross section, the cochlear duct is roughly triangular in shape and forms the central compartment of a larger bony canal that surrounds and conforms to its spiral trajectory. The medial edge of the cochlear duct is anchored to the osseus spiral lamina, a thin ridge of bone that protrudes laterally from the modiolus. The lateral wall of the cochlear duct is attached to the spiral ligament, a thickening of the periosteum that lines the bony cochlea.

Extending throughout much of the cochlear duct, along the line of attachment between the osseus spiral lamina and the modiolus, is the acoustic or spiral ganglion. Housed within an irregular cavity of modiolar bone, termed Rosenthal's canal, the spiral ganglion contains the cell bodies of auditory neurons which link the cochlear receptors to the central nervous system. The dendrites of these bipolar neurons reach their target receptors by first projecting through radial canals within the osseus spiral lamina and then emerging from the lateral margin of this bony shelf through small openings, the habenula perforata, to penetrate the overlying sensory epithelium. The axons of these neurons are grouped together in fascicles that exit from the medial edge of Rosenthal's canal, project through the core of the modiolus, and emerge from the bony cochlea to coalesce into the acoustic component of the VIIIth cranial nerve. All of these afferent fibers eventually synapse with first-order neurons located within the cochlear nuclei of the auditory brain-

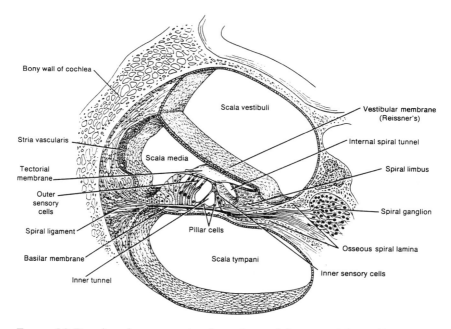

Figure 5.2. Drawing of a cross section through one of the turns of the cochlea. (From Bloom and Fawcett 1975).

stem (reviewed in Ryugo 1992). In addition to afferent neurons projecting from the cochlea to the brainstem, the spiral ganglion also contains the axons of efferent neurons projecting from nuclei within the superior olivary complex of the auditory brainstem to the sensory receptors within the cochlear duct (reviewed in Warr 1992).

The cochlear duct divides the bony cochlea into three fluid-filled compartments or scalae (Fig. 5.2). The enclosed fluid space within the cochlear duct is called the scala media and, like the rest of the membranous labyrinth, is filled with endolymph. In contrast, the upper and lower scalae (vestibuli and tympani, respectively) are osseus cavities that extend from the bony vestibule and, hence, are filled with perilymph. At the apex of the bony cochlea, the scalae vestibuli and tympani are joined by the helicotrema, a small opening that is formed when the interposed cochlear duct terminates as a blind sac, the cecum cupulare (see Fig. 5.1). At the base of the cochlea, the scalae vestibuli and tympani open, respectively, into the oval and round windows.

2.3 Anatomy of the Mammalian Cochlear Duct

Three histologically distinct tissues, Reissner's membrane, the basilar membrane, and the stria vascularis, comprise the walls of the cochlear duct and separate it from the surrounding perilymphatic scalae.

2.3.1 Reissner's Membrane

The roof of the cochlear duct is formed by the vestibular (Reissner's) membrane, a thin sheet of connective tissue bounded by two layers of epithelial cells. This membrane stretches across the cochlear canal, in an oblique fashion, from the spiral ligament lining the bony cochlear capsule to the spiral limbus, an upward bulging of the periosteum that lines the surface of the osseus spiral lamina. The cells in Reissner's membrane are joined together by tight junctions which form a fluid barrier that prevents mixing of the perilymph within the overlying scala vestibuli and the endolymph within the cochlear duct.

2.3.2 Basilar Membrane

The cochlear duct is separated from the underlying scala tympani by a floor of tissue, known as the basilar membrane, which extends horizontally across the cochlear canal from the bottom of osseus spiral lamina to the spiral ligament. This membrane supports the organ of Corti (the auditory sensory epithelium) and is composed of a homogeneous extracellular material (ground substance) containing slender (100 Å) filaments stretching across its width (Iurato 1962). The density of these filaments confers stiffness upon the basilar membrane; their orientation ensures that membrane stiffness is greater in the radial than the longitudinal direction (Voldrich 1978). In cross section, two distinct regions of the basilar membrane are recognizable: the arcuate and pectinate zones. The arcuate zone extends from the edge of the osseus spiral lamina to the feet of the outer pillar cells within the organ of Corti and contains a single layer of filaments surrounded by a small amount of ground substance. The pectinate zone extends from the outer pillar cells to the spiral ligament and contains a thicker layer of ground substance sandwiched between two layers of filaments. The undersurface of the entire basilar membrane is covered by a meshwork of mesothelial cells which form a tympanic cover layer.

In all species examined thus far, the total width of the basilar membrane increases systematically from the base to the apex of the cochlea. In many, but not all, species, the thickness of the basilar membrane decreases continuously from base to apex.

2.3.3 Stria Vascularis

The lateral wall of the cochlear duct is formed by the stria vascularis, an epithelium consisting of three cell layers surrounding a dense network of capillaries (for a review see Smith 1981). The luminal wall of the cochlear duct is composed, in part, by a sheet of dark cells within the inner margin of the stria. These marginal cells display a highly polarized morphology; their luminal faces, which are bathed in endolymph, are composed of straight, smooth membranes welded together by tight junctions. Their basolateral membranes are folded into numerous, slender, fingerlike processes that terminate freely within the stria and contact individual capillaries. Scattered among the marginal cells and capillaries are the intermediate cells which comprise the middle layer of the stria. In contrast to marginal cells, intermediate cells have a pale appearance due to a lower concentration of cytoplasmic organelles and possess only a few, short processes that end on capillaries. The outer layer of the stria vascularis, which faces the perilymph-filled spiral ligament, consists of flat rectangular basal cells interconnected through tight junctions.

In addition to providing a physical seal between the fluid spaces of the membranous and bony cochleae, the stria vascularis also actively maintains the ionic composition of the endolymph. Current research suggests that the marginal cells play a pivotal role in this process by transporting K^+ into and removing Na^+ from the scala media (for a recent review see Offner, Dallos, and Cheatham 1987). One consequence of this transport process is that the scala media, unlike most other fluid-filled spaces of the body, is maintained at a high positive potential (endocochlear potential, $EP^+ = 60$ to 100 mV) with respect to the blood (Békésy 1952).

2.4 Macromechanical Properties of the Cochlea: Sound-Induced Motion of the Cochlear Duct

The principal function of the membranous cochlea is to act as a hydromechanical frequency analyzer. Sound waves, collected from the external environment by the pinnae, propagate along the ear canal and impinge upon the eardrum, causing it to vibrate (see Rosowski, Chapter 6, for a discussion of the middle ear). The vibrations of the eardrum are conveyed to the fluid-filled cochlea by a series of three articulated bones that span the air-filled middle ear cavity and are suspended within it by minute connective tissue ligaments. The first bone in this ossicular chain is attached to and moves with the inner surface of the eardrum; the last bone, which resembles a saddle stirrup, is seated within the oval window recess of the bony vestibule where it moves in a pistonlike fashion against the flexible oval window membrane. Movements of the oval window produce pressure fluctuations within the adjacent fluid-filled scala vestibuli. As depicted in Figure 5.1, the scala vestibuli and scala tympani are joined through the helicotrema at the apex of the cochlea. The scala tympani, in turn, opens into a membranous round window at the base of the bony cochlea near its juncture to the vestibule. Consequently, inward motion of the oval window produces outward motion of the round window.

During extremely slow movements of the oval window, such as those occurring during static pressure changes within the middle ear or at very low (infrasonic) sound frequencies, displaced perilymph flows along the length of the cochlea and is freely exchanged between scalae vestibuli and tympani through the helicotrema. Under these conditions, the helicotrema acts like an acoustic shunt by reducing the pressure difference across the cochlear duct, which is the driving force for its displacement. (Parenthetically, it has been noted that for mammals of comparable body weight the size of the helicotrema is inversely related to cochlear sensitivity at low frequencies [Dallos 1970]. Presumably, as its radius increases, the helicotrema becomes more effective as an acoustic shunt.) At acoustic frequencies, the pressure wave originating at the oval window propagates throughout the scala vestibuli at the speed of sound in fluid (approximately 1400 meters/second). In the human cochlea, for instance, it takes only 25 μs for an acoustic pressure wave to propagate along the 35-mm length of the cochlear duct. The resulting pressure gradient which arises between the scalae vestibuli and tympani develops too quickly to be dissipated by the flow of perilymph through the helicotrema and therefore acts upon the intermediate cochlear partition to cause its displacement. The pattern of displacement exhibited by the mammalian cochlear duct takes the form of a traveling wave which has several unique features (Békésy 1960). First, this wave is unidirectional; it always originates at the base and travels towards the apex of the cochlea. Second, it exhibits a crest or peak of maximum displacement at a specific location along the cochlear duct that is directly determined by the frequency of the acoustic stimulus; high frequency stimulation causes maximum displacement near the base of the cochlear duct, lower frequencies at progressively more apical locations. This last feature of the traveling wave generates a cochlear frequency-place map which, in large measure, underlies the ability of mammals to analyze the spectra of impinging sounds.

The traveling wave of displacement exhibited by the cochlear duct is the consequence of a pronounced base-to-apex gradient in the stiffness of the basilar membrane (Békésy 1960). The comparative stiffness exhibited by the basilar membrane at a particular location along its length may be estimated by the ratio of its thickness to width at that point. In all mammals examined thus far, the basilar membrane widens progressively from base to apex, and, in most species, this increase in width is accompanied by a gradual decrease in thickness (see Table 5.1). Empirical measurements have suggested a 10^3 decrease in stiffness from base to apex (Békésy 1960).

3. Cellular Anatomy of the Auditory Sensory Epithelium

Acoustic transduction within the mammalian inner ear takes place within the organ of Corti, a complex epithelial ridge that protrudes into the scala media and rests on the basilar membrane throughout the length of the cochlear duct (Fig. 5.3).

The cells comprising the organ of Corti may be broadly divided into two categories: (1) cochlear hair cells, which are specialized sensory receptors that convert mechanical energy generated within the cochlear partition in re-

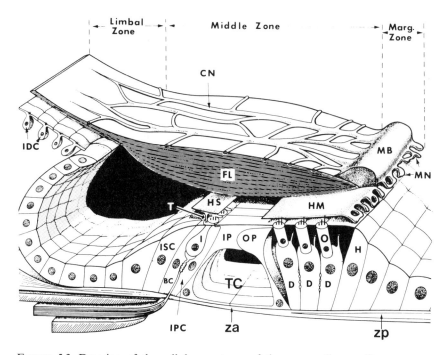

FIGURE 5.3. Drawing of the cellular anatomy of the mammalian auditory sensory epithelium (organ of Corti) as viewed in a radial cross section. Abbreviations: BC, border cell; CN, cover net of the tectorial membrane; D, Deiter's cell; FL, fibrous layer of tectorial membrane; MB, marginal band of the tectorial membrane; MN, marginal net of the tectorial membrane; H, Hensen's cell; HM, Hardesty's membrane; HS, Hensen's stripe; I, inner hair cell; IDC, interdental cell; IP, inner pillar cell; IPC, inner phalangeal cell; ISC, inner sulcus cell; O, outer hair cell; OP, outer pillar cell; T, trabecula; TC, tunnel of Corti; za, arcuate zone of basilar membrane; zp, pectinate zone of basilar membrane. (From Lim Archives of Otolaryngology, vol. 96 pp. 199–215 copyright 1972, American Medical Association).

sponse to sound into electrical signals that can be relayed, via chemical synapses, to the auditory brainstem; and (2) supporting cells, which are a morphologically diverse collection of electrically unexcitable cells that form a rigid, but flexible, framework surrounding and underpinning the auditory receptors.

Throughout the organ of Corti, cochlear hair cells and surrounding supporting cells are joined at their apices by tight junctions, forming a reticular lamina. The tight junctions within this lamina, together with those found within the lateral wall and roof of the cochlear duct, provide a fluid barrier which effectively seals off the endolymph-filled scala media from the surrounding perilymph-filled scalae. In contrast, the basilar membrane is relatively porous to the perilymph within the underlying scala tympani, and consequently, the extracellular spaces within the sensory epithelium are filled with this fluid and are equipotential with scalae tympani and vestibuli (Dallos 1985a). The cochlear hair cells and supporting cells are therefore chemically partitioned between endolymph at their apices and perilymph at their basolateral surfaces.

3.1 Cochlear Hair Cells

Sensory transduction within the mammalian cochlea is mediated by specialized columnar epithelial cells that are capped, on their apical surface, by several rows of modified microvilli or stereocilia. These bundles of stereocilia endow the receptors with a hirsute appearance, and consequently, they are commonly referred to as hair cells.

The sensory hairs of adult mammalian cochlear receptors are not true cilia. When viewed in cross section, they lack the usual array of nine peripheral and two centrally positioned microtubules that distinguish the motile cilia found within the respiratory, digestive, reproductive, and urinary systems and in the kinocilia of vestibular hair cells. Within each stereocilium, and extending throughout most of its length, is a large bundle of more than 3,000 actin filaments, each of which is extensively cross-linked to adjacent actin filaments through strands of fimbrin (DeRosier, Tilney, and Engelman 1980). At its insertion into the cuticular plate, each stereocilium undergoes a three- to five-fold reduction in diameter and generally contains less than 30 actin filaments (Tilney, DeRosier, and Mulroy 1980). This intrinsic organization confers a great rigidity on the stereocilium which, when deflected by fluid motion within the cochlear duct, does not flex but rather pivots about its base, much like a lever rotating about a fulcrum (Flock and Strelioff 1984).

The stereocilia of each mammalian auditory hair cell are organized into three to five rows oriented approximately parallel to the long axis of the cochlear duct (Engström, Ades, and Hawkins 1962). Each row contains stereocilia of approximately uniform height that are joined together by extensive lateral crossbridges. Adjacent rows are graded in height, much like a staircase, with the shortest row positioned closest to the modiolus. Between rows, the tips of individual shorter stereocilia are linked to their adjacent taller counterparts by slender upward-directed filaments. These "tip" links are ubiquitous among vertebrate hair cells (Pickles, Comis, and Osborne 1984), and their loss, due to mechanical or chemical trauma, renders the hair cell incapable of mechano-transduction (Hudspeth and Jacobs 1979; Assad, Shepherd, and Corey 1991).

In all mammals examined so far, two classes of cochlear auditory hair cells may be readily distinguished. The nomenclature used to describe these receptors reflects their different spatial locations within the organ of Corti; those positioned closest to the osseus spiral lamina are referred to as inner hair cells, whereas those located more peripherally within the sensory epithelium, closer to the lateral wall of the cochlear duct, are termed outer hair cells. Both inner and outer hair cells are organized into precise longitudinal rows which extend almost the entire length of the cochlear duct. Apart from the spatial segregation of inner and outer hair cells, these receptors are further distinguished by: (1) their cellular morphology; (2) their physical coupling to adjacent supporting cells within the organ of Corti; and (3) their pattern of innervation.

3.1.1 Inner Hair Cells

In the majority of mammals examined, inner hair cell receptors form a single row oriented parallel to the cochlear spiral. The number of inner hair cells found within an individual animal depends upon the length of its cochlear duct. The mouse cochlea, which is 6.8 mm in length, contains 765 inner hair cells (Ehret and Frankenreiter 1977), whereas the human and dolphin cochleae, averaging 35 and 38.5 mm in length, respectively, each house approximately 3500 (Nadol 1988; Wever et al. 1971c).

The inner hair cell is shaped like a flask with a rounded cell soma tapering into a thinner, elongated neck. The spherical nucleus, positioned midway along the length of the inner hair cell, divides this receptor into two topographic domains. Below the nucleus lies the basal, or neural, pole that receives numerous synaptic contacts from up to twenty or more afferent cochlear nerve fibers. In all mammals examined to date, inner hair cells receive the bulk (90 to 95%) of all afferent contacts with cochlear nerve fibers (see Ryugo 1992). Above the nucleus, the inner hair cell membrane is synapse free and terminates apically in a flattened, elliptical cuticular plate crowned by a forest of stereocilia in nearly straight rows. A detailed description of the ultrastructure of cochlear hair cells may be found in Lim (1986).

Inner hair cells vary in length from 30 to 40 μm, depending upon the species. For an individual animal, however, cell size does not vary significantly along the cochlear spiral. In contrast, the size and number of stereocilia on each inner hair cell varies with both the species examined and the cochlear location sampled. In all species examined thus far, the length and number of inner hair cell stereocilia increase modestly from the base to the apex of the cochlea (Lim 1980; Saunders and Garfinkle 1983; Wright 1984; Dannhof and Bruns 1991; Vater and Lenoir 1992). For quantitative data on this topic, see Table 5.2.

3.1.2 Outer Hair Cells

The outer hair cell receptors comprise 75 to 80% of the sensory cell population within the mammalian cochlea. They are organized into three to five parallel rows extending along the length of the cochlear duct, and as for inner hair cells, their number depends upon cochlear size. The mouse ear contains approximately 2,526 outer hair cells (Ehret and Frankenreiter 1977), the human ear 12,000 (Nadol 1988), and the dolphin ear 13,933 (Wever et al. 1971c). Although outer hair cell receptors receive a sparse (5 to 10%) afferent inner-

TABLE 5.2. Comparative anatomy of cochlear hair cells.

Species	Body weight (kg)	Audible frequency range (kHz)	Length of inner hair cell soma (μm)	Length of inner hair cell cilia (μm)	Number of cilia per inner hair cell	Length of outer hair cell soma (μm)	Length of outer hair cell cilia (μm)	Number of cilia per outer hair cell	Authors
mole rat	0.08	0.1 to 10.0				base: 30 apex: 75	base: 3.5 apex: 12	base: 120 apex: 30–40	Bruns et al. 1988
human	75	0.13 to 16.0	base: 31 apex: 34	base: 4.5 apex: 5.5	base: 41 middle: 64	base: 28 apex: 66	base: 4.5 apex: 8.0	base: 120–148 apex: 46–80	Nadol 1988
chinchilla	0.8	0.09 to 25.0	base: 35 apex: 35	base: 1.75 apex: 4.75	base: 75 apex: 65	base: 25 apex: 45	base: 1.0 apex: 5.5	base: 80–110 apex: 40–90	Lim 1980
guinea pig	0.1	0.2 to 45.0				base: 21 apex: 80	base: 1.2 apex: 5.8		Wright 1984
cat	2.5	0.125 to 60.0			base: 60	base: 20 apex: 50	base: 2.0 apex: 6.0	base: 100–120	Spoendlin 1966
rat	0.2	1.0 to 59.0		base: 2.3 apex: 4.3	base: 32 apex: 41	base: 15 apex: 40	base: 2.7 apex: 4.4	base 75 apex: 62	Iurato 1962
bat	0.02	1.0 to 200.0	base: 30 apex: 40	base: 2.5 apex: 7.0		base: 8 apex: 15	base: 1.0 apex: 2.5		Dannhof and Bruns 1991

vation from the cochlear nerve, they are contacted by numerous efferent nerve fibers originating from the superior olivary complex within the auditory brainstem (Warr 1992).

The mammalian outer hair cell is cylindrical in shape with a large spherical nucleus positioned close to the basal (neural) pole of the cell. The most distinguishing cytological feature of the outer hair cell is the presence of multiple layers of subsurface cisterns composed of lamellar endoplasmic reticulum that extend along the long axis of the cell from the level of the nucleus to the cuticular plate (for a review see Lim 1986). Extending from the cuticular plate are three rows of stereocilia that are organized into a W configuration at the base of the cochlea and a U configuration at the apex (Engström, Ades, and Hawkins 1962).

The size of the outer hair cell soma varies greatly with species and cochlear location (Retzius 1884; Wada 1923; Pujol et al. 1992). Within the organ of Corti, hair cell length increases with both radial and longitudinal position (Bohne and Carr 1979; Lim 1986). At a given cochlear location, outer hair cells positioned closest to the modiolus are shorter than those located more laterally within the sensory epithelium (Fig. 5.3); along the length of the cochlea, the outer hair cells at the apex are, on average, twice as long as those positioned at the base (Table 5.2). Outer hair cell stereocilia length and number also change with longitudinal location; at the apex of the cochlea, each outer hair cell contains fewer but longer stereocilia than do those at the base (Bruns 1979; Bruns and Goldbach 1980; Bruns et al. 1988; Saunders and Garfinkle 1983; Wright 1984; Lim 1986; Dannhof and Bruns 1991; Vater and Lenoir 1992). (Table 5.2).

3.1.3 Supporting Cells

In a radial view of the organ of Corti (Fig. 5.3), the most conspicuous elements are the centrally positioned inner and outer pillar cells which extend throughout the length of the cochlear duct and form the tunnel of Corti. Resting directly on the basilar membrane, these cells have adjoining, widely spaced conical bases that taper upward into slender cylindrical processes that expand and interlock at the surface of the organ of Corti. These cellular "pillars" contain an extensive network of microtubules that confer upon them a great rigidity.

Located medially to the inner pillar cells, the inner hair cell receptors are firmly encased within a matrix of supporting cells that lies above the rigid osseus spiral lamina. Border cells contact the medial (modiolar) side of the inner hair cell, inner phalangeal cells flank its sides, and inner pillar cells abut its lateral surface. This structural arrangement strongly suggests that inner hair cell receptors are minimally displaced during direct movements of the cochlear duct (Lim 1986).

Positioned laterally to the outer pillar cells are the outer hair cell receptors which are held in place by the Deiter's cell, a sort of cellular C-clamp. The base of the Deiter's cell rests on the pectinate zone of the basilar membrane with its apex exhibiting two specializations: (1) an invagination which forms a pocket that encases the basal third of the outer hair cell and its neural supply; and (2) an evagination in the form of a slender rigid fingerlike process that projects away from the soma of the outer hair cell and extends to the surface of the organ of Corti where its expands into a phalangeal process anchoring the apex of the outer hair cell at the level of the cuticular plate. Consequently, although yoked at its apical and basal poles, two-thirds of the outer hair cell remains free of cellular attachment. The location of the outer hair cell receptors within the organ of Corti, and their means of attachment to it, places them in ideal positions to be influenced by or to act upon the sound-induced movements of the basilar membrane (Lim 1986).

3.1.4 Tectorial Membrane

The tectorial membrane is a large acellular flap of tissue positioned above the organ of Corti and extending along most of its length (Fig. 5.3). In a cross-section of the cochlear duct, three topographically distinct regions of the tectorial membrane have been recognized (Lim 1972): (1) a medial or limbal zone, comprised of a thin rim of tissue attached to the surface of the spiral limbus; (2) a thick central or middle zone overlying the sensory cells and containing radial, unbranched protofibrils embedded within an amorphous ground substance; and (3) a lateral or marginal zone that terminates in a marginal band (cochlear base) or marginal net (cochlear apex) and anchors the peripheral edge of the tectorial membrane to the supporting cells of Hensen and Deiter. The surface of the tectorial membrane is covered by a net of coiled, branched fibrils.

The bottom of the tectorial membrane displays two specializations associated with the underlying sensory receptors. Laterally is a thickening referred to as Hardesty's membrane, but originally observed by Kimura (1966), that attaches directly to the tallest stereocilia of the three rows of outer hair cells (Lim 1972). Above the inner hair cells, the tectorial membrane forms another thickening, Hensen's stripe, but direct attachment between this structure and the inner hair cell stereocilia has been observed only within the base of the cochlea in a few species (Hoshino 1976). The general view derived from ultrastructural studies is that the outer hair cell stereocilia are firmly attached to the undersurface of the tectorial membrane, but the inner hair cell stereocilia are freestanding (Lim 1986).

3.2 Micromechanical Properties of the Cochlea: Motion Within the Organ of Corti

The gross dimensions of the cochlea determine that the frequency components of a sound are sorted and mapped along the length of the basilar membrane and its overlying organ of Corti (Békésy 1960). The location of the peak of the cochlear excitation pattern caused by a single tone is a monotonic function of the tone's frequency, with low frequencies represented at the apex and higher frequencies represented progressively more basalward. This cochlear frequency map is determined to a great extent by a progressive increase in the stiffness of the cochlear partition from apex to base. Stiffness is highly correlated with the basilar membrane's decreasing width and increasing thickness from apex to base (Békésy 1960). This essential view is generally accepted today (e.g., Dallos 1992), and the frequency-place maps Békésy (1960) determined on the radically dissected cochleae of several mammalian species remain generally valid. However, research during the last 25 years has revealed that mechanical frequency selectivity and sensitivity depend on the physiological integrity of the cochlea (for a review see Rhode 1984). It is now generally presumed that the relative motion patterns of cells and accessory structures within the organ of Corti (micromechanics) and the physiological properties of these cells determine the sensitivity and frequency selectivity of the mechanical response of the basilar membrane (Dallos 1992). The following sections discuss both the passive and active (physiologically determined) mechanisms that normally contribute to the acute frequency selectivity and sensitivity of the mammalian cochlea. For a detailed and fascinating historical perspective on the key experiments in auditory neurobiology which have led to the concept of an "active" cochlea, the reader is referred to an excellent review by Dallos (1988a).

3.2.1 Passive Mechanisms

Békésy (1960) originally considered the tectorial membrane to be a relatively stiff, platelike structure to which the stereocilia of both inner and outer hair cells were attached. In this view, shearing of the stereocilia is a simple function of the amplitude of basilar membrane motion. Zwislocki (1986) has now found that the tectorial membrane in vivo is more gel-like than Békésy believed and that it may have passive resonance-like properties that interact with those of the basilar membrane to cause hair cell stimulation. These interactions of two tuned filters are thought to enhance the mechanical sensitivity and frequency selectivity of the cochlea (Zwislocki and Kletsky 1979). Since the structural properties of the tectorial membrane are profoundly altered when the ionic composition of the surrounding endolymph is changed (Kronester-Frei 1979), it is likely that Békésy, by opening the cochlear duct and thereby allowing the replacement of endolymph by perilymph, would not have been able to observe these effects.

With regard to the possible role of the tectorial membrane in the enhancement of cochlear function, it is interesting to note that a correlation has been observed between the mass of this structure and regions of the cochlea exhibiting enhanced sensitivity. The mustache bat, *Pteronotus parnelli*, has a pronounced thickening of the tectorial membrane at those basal cochlear locations that exhibit pronounced frequency selectivity and threshold sensitivity for 30 kHz, 60 kHz, and 90 kHz, the principal frequency components of its biosonar signal (Henson and Henson 1991). In contrast, the frog-eating bat, *Trachops cirrhosus*, which is sensitive to the relatively low frequency calls (less than 5 kHz) of its prey, displays a pronounced enlargement of the tectorial membrane within the apical cochlea (Bruns, Burda, and Ryan 1989). A similar enlargement of the tectorial membrane is also present within the apical cochlea of the mole rat, *Spalax ehrenbergi*, an animal limited to low frequency hearing (Bruns et al. 1988). Finally, Bruns, Burda, and Ryan (1989) have noted that in the gerbil, *Pachyuromys duprasi*, the frequency of greatest sensitivity (1.5 kHz) is also represented at the cochlear region where the tectorial membrane reaches maximum mass.

Pujol and colleagues (1992) have compiled data from a wide range of species on the comparative differences in the lengths of outer hair cells and their stereocilia as a function of cochlear location. Figure 5.4 shows their observations. In general, longer stereocilia and cell bodies are found at the apex of the cochlea where the lower frequencies are represented. Inner hair cells, on the other hand, are much more structurally homogeneous throughout the cochlea. In addition, the longest outer hair cells and stereocilia are

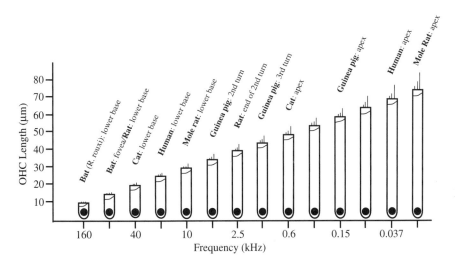

FIGURE 5.4. A schematic illustration of the relationship between the size of an outer hair cell and the frequency of best response exhibited at its location within the cochlea. Note that for each species depicted the length of both the outer hair cell soma and its stereocilia decreases progressively as the cochlear location shifts toward the base and the upper frequency limit of hearing rises. Species and cochlear locations sampled are depicted above the hair cell; best frequency at that particular cochlear location is provided below and was derived from cochlear frequency-place maps available for that species. (Reprinted from Auditory Physiology and Perception, Advances in the Biological Sciences, Vol. 83 pp. 45–52, copyright 1992, with kind permission from Pergamon Press Ltd., Headington Hall, Oxford OX3 OBW, U.K).

found in species adapted for hearing well at low frequencies (e.g., mole rat, human, and guinea pig) while the shortest are found among species that hear well at relatively high frequencies (e.g., bat and rat). These observations show that the outer hair cells have gross structural properties that may be adaptations for frequency selectivity within particular frequency regions.

3.2.2 The Active Cochlea: Outer Hair Cell Motility

The major concept to emerge in hearing science over the past 15 years is that the cochlea is active, providing energy of its own that interacts with the passively conducted acoustic energy to enhance mechanical sensitivity and frequency selectivity (e.g., as reviewed by Dallos 1992). Recent in vitro studies of isolated cochlear receptors have shown that the outer hair cells undergo reversible length changes in response to electrical stimulation (Brownell et al. 1985; Ashmore 1987), neurochemical stimulation (Brownell et al. 1985), and acoustic (mechanical) stimulation (Canlon, Brundin, and Flock 1988). In addition, this acoustic response shows intrinsic frequency selectivity that varies among outer hair cells with their location along the cochlear duct (Brundin, Flock, and Canlon 1989; Brundin et al. 1992). In these studies, the lengths of isolated outer hair cells were measured during mechanical stimulation at frequencies between 200 and 10,000 Hz. Each hair cell's motile response to stimulation was highly frequency selective, and the most effective frequency was correlated with hair cell length. Short cells from the basal region of the cochlea responded best to high frequencies, and long cells from the apex responded best to low frequencies. The implication from these findings and from studies of electromotility (Kachar et al. 1986; Dallos, Evans, and Hallworth 1991) is that the outer hair cells function as motors, controlled by the acoustic input, that somehow alter local micromechanics to amplify the sensitivity and frequency selectivity of the cochlea. The inner hair cells detect the resultant enhanced motion of the basilar membrane and transmit this information to the central nervous system through their copious afferent innervation.

In addition to relatively slow changes in length, isolated outer hair cells have been observed to oscillate in length (3 to 5%) at the frequency of stimulation (Kachar et al. 1986; Ashmore 1987). This suggests that both sustained and oscillatory length changes may play a role in selectively amplifying the traveling wave described by Békésy (1960). Neither slow nor oscillatory motility has been observed in inner hair cells or in the inner ears of nonmammals.

4. Longitudinal Variations in Cochlear Structure: Two Cochleae in One?

There is growing evidence that the basal (high-frequency) and apical (low-frequency) ends of the cochlea function somewhat differently (Dallos 1988b). For example, apical outer hair cells are electrically activated by low to moderate sound levels (Dallos 1985b), whereas basal outer hair cells are excited only by intense sounds (Cody and Russell 1987). In addition, the tuning curves of primary auditory afferents from the apical regions of the mammalian cochlea are reminiscent of those obtained from submammalian vertebrates and tend to be simple, symmetrical, and "V-shaped," while those obtained from the base tend to be more complex and asymmetrical with prominent low-frequency "tails" (Kiang et al. 1965). In most mammals there are also conspicuous structural differences in apical and basal outer hair cells. For example, the efferent innervation to outer hair cells is relatively dense in the base and rather sparse in the apex (Spoendlin 1966). In addition, there are pronounced differences in the lengths of outer hair cells (Retzius 1884; Wada 1923; Bohne and Carr 1979; Pujol et al. 1992), the height of their stereocilia (Bruns 1979; Bruns and Goldbach 1980; Saunders and Garfinkle 1983; Wright 1984; Lim 1986; Dannhof and Bruns 1991; Vater and Lenoir 1992), and the organization of their cytoskeleton (Thorne et al. 1987) between the cochlear base and apex. These differences suggest that the structures required for processing low frequencies may be somewhat different from those required for processing high frequencies. For example, precise temporal synchronization between the acoustic wave-form and the neural response (i.e., phase locking) may be a useful representation of frequency in apical regions, but not at the most basal regions (Rose et al. 1967). At the same time, frequency is represented well by the profile of neural activity across tonotopic arrays of cochlear afferents from basal regions (Young and Sachs 1979), but not at the most apical regions. It would not be surprising to find adaptations for accurate temporal coding at the apex and for accurate spatial coding at the base.

Low-frequency hearing is a primitive vertebrate characteristic, while high-frequency hearing (above 10 kHz) is found only among the mammals (Fay 1988). It might be expected, then, that the cochlear apex would show some primitive vertebrate characteristics while the base would show specialized mammalian characteristics. It could also be expected that mammals may differ with respect to the frequency range and cochlear location at which these different types of adaptations occur.

5. Species Variation in Cochlear Structure and Function: Cochlear Maps

Based on a limited sample of mammalian species for which cochlear structure and function have been adequately studied, we propose a distinction between mammals having a generalized cochlea (the generalists) and those having morphological specializations for processing sound in particular frequency regions (the specialists). This distinction is made primarily on the basis of cochlear frequency-place maps illustrated in Figure 5.5.

Figure 5.5 shows empirical cochlear maps for the cat (Liberman 1982), rat (Müller 1991), opossum (Müller, Wess, and Bruns 1993), mole rat (Müller et

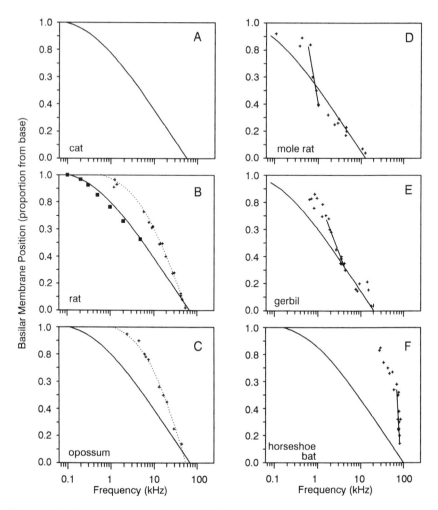

FIGURE 5.5. Empirical (+ symbols and filled squares) and theoretical (continuous lines) cochlear maps for six mammals. For the + symbols, the location of the peak of the cochlear excitation pattern was estimated from the location within the cochlea exhibiting maximum labeling following the injection of HRP intracellularly into cochlear fibers (cat; panel A) or extracellularly into the cochlear nucleus (all other panels). The frequency plotted is the frequency with the lowest threshold for single cochlear nerve units (panel A) or for unit clusters of the cochlear nucleus (all other panels). The filled squares for the rat (panel B) are from the direct observations of basilar membrane displacement patterns by Békésy (1960). The solid lines were fit with Equation 1 with the slope parameter set to 2.1 and parameter A adjusted to force the function to zero on the ordinate at the upper frequency limit of hearing (see text). The dashed lines were fit by eye with Equation 1 by allowing the slope parameter (1.0 for panel B and 0.75 for panel C) and parameter A to vary. The thin, straight lines in panels D, E, and F were drawn to indicate the regions of steepest slopes. See text and Table 1 for further details and references.

al. 1992), gerbil (Müller, Ott, and Bruns 1991), and horseshoe bat (Vater, Feng, and Betz 1985). The data for the cat were obtained by injecting horseradish peroxidase (HRP) into physiologically identified cochlear nerve fibers and tracing them to their point of innervation in the cochlea. All other maps shown were determined by injecting HRP into small, physiologically characterized regions of the cochlear nucleus.

Data for the cat were fit by Liberman (1982) with a function developed by Greenwood (1961, 1962, 1990) that has the following form:

$$(1) \quad f_{Hz} = A(10^{(2.1p)} - 0.8) \tag{1}$$

where p is a position as a proportion of cochlear length (from the apex), f is the frequency in Hz, and A is a free parameter. Greenwood (1990) has pointed out that the slope parameter of 2.1 can be used to model all the species he investigated if the cochlear length is normalized to unity. This means that only one additional parameter, the species-specific constant A, is required to estimate the normalized cochlear position-frequency map. It is significant that Equation 1 is of the same form and slope as the function relating basilar membrane elasticity to the distance from the apex as determined by Békésy (1960) for the elephant, human, guinea pig, rat, and mouse. In Figure 5.5A, the fit between Liberman's (1982) data and Equation 1 is so good that only the model function is plotted. It should be pointed out that Equation 1 also accounts very well for direct and indirect estimates of the cochlear map in humans (Greenwood 1990) and for Békésy's (1960) maps for the rat (Fig. 5.5B), elephant, guinea pig, and mouse as determined by direct observation (reviewed by Fay 1992).

Fay (1992) determined that the parameter A can be estimated from independent measures of the upper frequency range of hearing (at 60 dB Sound Pressure Level (SPL)) for the species studied by Békésy (1960). Using estimates of the upper hearing range for the species represented in Figure 5.5, the maps according to Equation 1 are plotted as solid lines. Clearly, the maps estimated for the rat, opossum, mole rat, gerbil, and horseshoe bat do not fit the data well. On the other hand, the data for the rat and opossum are fit quite well by Equation 1 by adjusting parameter A and the slope parameter (dotted lines). In other words, an equation of the same form as Equation 1 can fit the data for the rat and opossum. In this sense, these cochleae can be considered scale models of one another. This is not the case for the species to the right side in Figure 5.5. It is obvious, by inspection, that the data for the mole rat, gerbil, and horseshoe bat cannot be well fit by Equation 1; they all show abrupt slope transitions uncharacteristic of Equation 1. This is clearest for the horseshoe bat and mole rat, and less clear for the gerbil.

This analysis leads to defining generalized cochleae as like those on the left of Figure 5.5 in which the cochlear map data can be fit by adjusting the parameters of Equation 1. Specialized cochleae are defined as like those on the right of Figure 5.5, having empirical maps that cannot be reasonably fit by Equation 1.

It should be noted that all of the data in Figure 5.5, except for the cat, deviate from the theoretical predictions of Greenwood (1990) in the same direction; the data have steeper slopes than the theoretical functions. Similarly, all of the deviant data were determined using the same method; HRP injections in small regions of the cochlear nucleus. This leaves open the possibility that the mapping procedures based on CNS injections tend to overestimate the slopes of the cochlear maps. In this context, note that in Figure 5.5B the map for the rat, as determined by Békésy (1960) using direct observation (filled squares), corresponds more closely to the Greenwood function with a slope of 2.1 than do the more recent data (crosses). It is not clear whether the slopes of cochlear maps were underestimated in the postmortem preparations of Békésy (1960, cf. Müller 1991) or were overestimated by the use of HRP injections within the cochlear nucleus.

5.1 Generalized vs. Specialized Cochleae

The cochleae of the mole rat and horseshoe bat have been defined as specialized based on regions of abrupt transitions and steep slopes within their respective cochlear maps, whereas the rat has been categorized as possessing a generalized cochlea. What anatomical features might be associated with these differences in cochlear frequency mapping, and what are the consequences of such features for hearing?

The specialized frequency maps for the horseshoe bat and mole rat are associated with morphological features of the basilar membrane that are unusual (Fig. 5.6). The basilar membrane of the horseshoe bat, for example, is exceedingly thick (upper panel) and relatively wide (middle panel) within the first 20 to 30% of its length from the base. The middle region of the cochlea (30 to 70% from the base) contains a secondary spiral lamina but is otherwise unremarkable in structure. What is remarkable about the basilar membrane (BM) in this segment of the cochlea, however, is that it has a relatively constant thickness (about 8 μm in upper panel) and width (about 90 μm in middle panel). The stiffness of the BM throughout the middle region of the cochlea (estimated by the ratio of thickness to width) thus remains nearly constant (bottom panel). Since the filtering characteristics of the cochlea depend on mass and stiffness (Békésy 1960), the middle region (about 40% of the cochlea) responds to a very restricted range of frequencies. The function of the very thick region of the basal BM is not clear, but its edge clearly limits the extent of the middle, "constant stiffness" portion of the BM. Thus, in the horseshoe bat, the cochlear specialization primarily responsible for the steep portion of the cochlear map is the very shallow gradient of stiffness in the middle region.

The functional significance of this specialization is relatively well understood. The horseshoe bat echolocates using a biosonar signal consisting of a constant frequency (CF) portion terminated by a downward, frequency-modulated sweep (reviewed in Pollak 1992). The CF component is used to

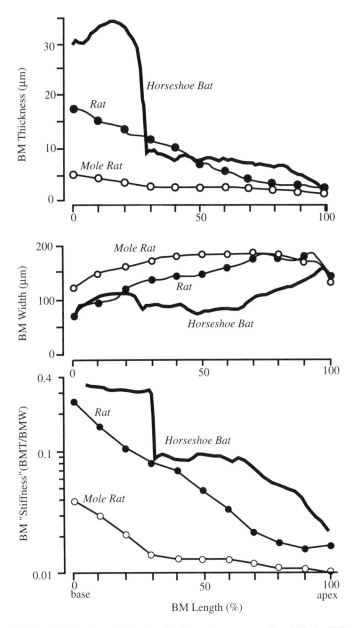

FIGURE 5.6. Longitudinal variations in thickness (upper panel), width (middle panel), and computed stiffness (lower panel) of the basilar membrane in the horseshoe bat (data from Bruns 1976), rat (data from Burda, Ballast, and Bruns 1988), and mole rat (data from Müller et al. 1992).

detect Doppler shifts in the returning echoes due to relative movements between the bat and its prey. One obvious function of the constant stiffness region of the cochlea is in frequency resolution. Since a small change in frequency results in a relatively large shift in the peak of the excitation pattern on the BM, frequency resolution is increased (Long 1977, 1980). Thus, the horseshoe bat is specialized for acute frequency analysis within a narrow frequency range (near 80 kHz) by virtue of a relatively large region of the cochlea having a very shallow stiffness gradient (Bruns 1976).

The cochlear map for the mole rat is similar to that of the horseshoe bat in having an especially steep portion of the cochlear map at middle portions of the cochlea that have a nearly flat stiffness gradient (Fig. 5.6, bottom panel). The mole rat differs from the horseshoe bat in that the frequencies mapped to this region are in the 0.6 to 1 kHz range, rather than at 80 kHz. It seems likely that frequency resolution is especially enhanced in this range, but there are no psychophysical data on this point. Unlike the horseshoe bat, the behavioral audiogram for the mole rat shows no unusual features except, perhaps, that sensitivity and upper frequency range is poor by mammalian standards (Heffner and Heffner 1992). The functional significance of the mole rat's apparently enhanced frequency resolution in the 0.6 to 1 kHz range has yet to be demonstrated.

The cochlear map for the gerbil (Fig. 5.5E) appears to have a relatively steep slope in its middle section (1.5 to 4 kHz). The cochlea of this species also has some unusual features (e.g., Webster and Webster 1977, 1980; Plassmann, Peetz, and Schmidt 1987). Compared with the rat, for example, the cochlea of the gerbil is relatively long (11.7 mm) and wide (350 μm) and has an obvious hyaline thickening of the basilar membrane in the apical region. In combination with its unusually large middle ear volume (see Rosowski, Chapter 6) and good hearing sensitivity at low frequencies (Ryan 1976), these cochlear features have been interpreted as special adaptations for processing low-frequency sound (Lay 1972; Bruns 1979; Plassmann, Peetz, and Schmidt 1987). The steep portion of the gerbil cochlear map resembles those of the mole rat and horseshoe bat, and the basilar membrane within this region exhibits a plateau in width (300 μm) and a peak in thickness (45 μm; Plassmann, Peetz, and Schmidt 1987), but stiffness estimates have not been made.

In contrast to mammals exhibiting plateaus and abrupt transitions within their cochlear frequency-place maps, those species exhibiting frequency-place maps well fit by Greenwood's function, such as the rat (Fig. 5.6), possess cochleae in which both BM thickness and width vary continuously from base to apex (Burda, Ballast, and Bruns 1988). Additional examples of such species include mice (Ehret and Frankenreiter 1977; Burda, Ballast, and Bruns 1988) and guinea pigs (Fernández 1952).

The data of Figures 5.5 and 5.6 begin a definition of cochlear generalists and specialists among the mammals based upon basilar membrane structure-function relationships. In addition, it has been well documented that species having specialized cochlear maps (i.e., those with abrupt transitions in slope) also often have unusual structural features associated with the organ of Corti (Bruns 1979).

Figure 5.7 illustrates the comparative morphology of the organ of Corti in several general and specialized mammals. Plates A through D depict the basal half of the cochlea and plates E through H depict the apical half. In each montage, the upper two illustrations illustrate the generalized cochlear morphology observed in humans (A and E) and cats (B and F), whereas the lower two illustrations represent mammals specialized for either high-frequency hearing, the horseshoe bat (C) and dolphin (D), or low-frequency hearing, the gerbil (G) and mole rat (H).

In species with especially well developed high-frequency hearing (C and D) the basal cochlea exhibits some features not found in more generalized mammals (A and B). (1) The basilar membrane is relatively narrow and thick. (2) A prominent secondary spiral lamina anchors the basilar membrane to the lateral wall of the cochlear duct. This lamina is found throughout the cochlear duct in mammals with high-frequency hearing, such as shrews and bats, but is present only in the basal cochlear region in mammals with low-frequency hearing, such as gerbils, guinea pigs, man and moles (Bruns, Burda, and Ryan 1989). (3) The supporting cells positioned laterally within the organ of Corti, the cells of Hensen and Claudius, are greatly hypertrophied. (4) The cochlear receptors and the extracellular spaces surrounding them are reduced in size (Bruns 1979).

Structural modifications are also found within the apical half of the cochlea in some species of gerbils (G) and in certain heteromyid rodents such as the kangaroo rat (not illustrated). These include: a prominent thickening of the pectinate zone of the basilar membrane, termed a hyaline mass and greatly hypertrophied Hensen's cells (Lay 1972; Webster and Webster 1977, 1980; Bruns 1979; Plassmann, Peetz, and Schmidt 1987). The apical half of the mole rat cochlea (H) also exhibits several specializations when compared with generalized species (E and F). (1) The scalae are particularly narrow. (2) The tectorial membrane is enlarged. (3) The outer hair cells are especially large, arrayed in up to six rows, and not separated by extracellular spaces (Bruns et al. 1988; Raphael et al. 1991)

It should be pointed out that regions of the cochlea that exhibit these unusual anatomical features do not necessarily overlap perfectly with those cochlear locations associated with unusual frequency maps. In the horseshoe bat and mole rat, at least, cochlear map specializations simply occur where there are very shallow gradients of stiffness.

Cochlear map specializations may enhance frequency resolution at high- (horseshoe bat), mid- (gerbils) or low- (mole rat) frequency regions. At the same time, generalized cochlear maps may be found in species that hear particularly well at high (e.g., rat and mouse), middle (e.g., cat), and low (e.g., human) frequencies. Thus, specialized cochlear maps are probably best viewed as adaptations for frequency analysis within particular bands rather than as adaptations for especially high- or low-frequency hearing per se. This

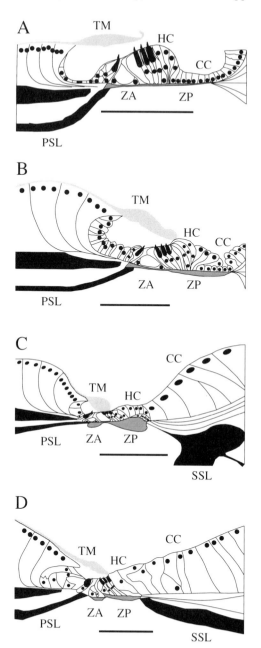

FIGURE 5.7. Illustrations of the organ of Corti from the basal (A-D) and apical (E-H) regions in mammals with generalized cochleae, human (A and E) and cat (B and F), and in mammals exhibiting cochlear specializations for processing high frequencies, horseshoe bat (C) and dolphin (D), and low frequencies, gerbil (G) and mole rat (H). All scale bars equal 100 μm. Cochlear receptors are highlighted in black. Abbrevia-

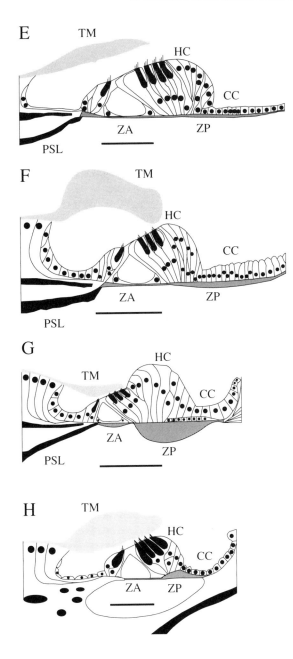

tions: CC, Claudius cells; HC, Hensen's cells; PSL, primary osseus spiral lamina; TM, tectorial membrane; SSL, secondary osseus spiral lamina; ZA, arcuate zone of basilar membrane; ZP, pectinate zone of basilar membrane. Human and cat (after Retzius 1884); gerbil (after Bruns 1979); mole rat (after Bruns et al. 1988); horseshoe bat (after Bruns 1980); dolphin (after Wever et al. 1971a).

suggestion can be evaluated using behavioral measures of frequency analysis for selected generalized and specialized species (e.g., as reviewed in Fay 1988).

5.2 What Can Be Predicted About Hearing from Gross Morphology of the Cochlea?

In general, an ability to accurately predict hearing capabilities from readily observable gross features of cochlear morphology would be valuable to the comparative and evolutionary study of hearing and to the selection and development of animal models for aspects of human hearing. In this section, we discuss the extent to which we can and cannot make simple predictions about hearing from cochlear morphology. All of the data plotted in the following figures were taken from Table 5.1.

Figure 5.8 plots a species' hearing range in octaves (at 30 dB above threshold) vs. the number of turns of its cochlea. Overall, hearing range correlates only weakly with the number of cochlear turns. Based primarily on data from species with generalized cochleae (open circles), West (1985) and Plassman, Peetz, and Schmidt (1987) concluded that there is a useful correlation between hearing range and the number of cochlear turns. However, Figure 5.8 shows that this correlation is considerably reduced when specialized species (filled squares) are included in the sample (e.g., horseshoe bat, mole rat, and dolphin). Without independent knowledge of possible cochlear specializations, predictions of hearing range from the number of cochlear turns will be of little value.

Figure 5.9 shows the relationship between the low-frequency limit of hear-

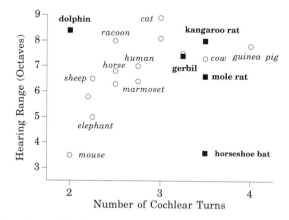

FIGURE 5.8. The relationship between the number of cochlear turns and hearing range in octaves is plotted for 19 species of mammals. All data in Figures 5.8 through 5.12 are provided in Table 5.1. In each figure, solid symbols represent mammals possessing specialized cochleas, open symbols are those with generalized cochleae. For purposes of clarity, not all species are labeled.

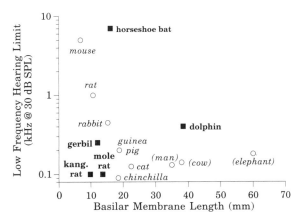

FIGURE 5.9. The relationship between basilar membrane length and low frequency hearing limit is depicted for 13 mammals with specialized and generalized cochleae.

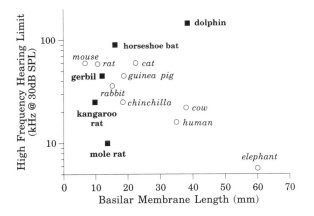

FIGURE 5.10. The relationship between basilar membrane length and high frequency hearing limit is illustrated for 14 mammals with specialized and generalized cochleae.

ing (at 30 dB above threshold) and basilar membrane length. The correlation between these variables is robust for species with generalized cochleae (open circles; West 1985), but declines when species with specialized cochleae are included (filled squares). Again, independent knowledge about whether or not a species' cochlea is specialized is required to be able to use basilar membrane length to predict the limit of low-frequency hearing. For generalists, when large-headed mammals are excluded (human, cow, elephant), as cochlear length decreases, the lower frequency limit of hearing progressively rises.

The high-frequency limit of hearing (at 30 dB SPL) is plotted vs. basilar membrane length in Figure 5.10. For specialized species (filled squares), there

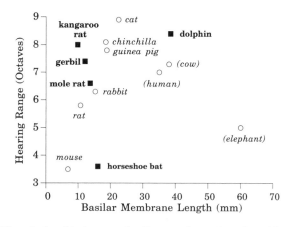

FIGURE 5.11. The relationship between basilar membrane length and hearing range in octaves is depicted for 14 mammals with specialized and generalized cochleae.

appears to be no useful relationship between these variables. For generalized species (open circles), however, as basilar membrane length decreases, the high-frequency limit of hearing progressively rises. Considered together, the data plotted in Figures 5.9 and 5.10 suggest that, in generalized species, as basilar membrane length decreases, the range of hearing shifts to higher frequencies.

Figure 5.11 shows the relationship between hearing range in octaves and basilar membrane length. Overall, there is a weak positive correlation such that species with longer cochleae tend to have a wider frequency range of hearing. If the analysis is restricted to relatively small, ground-dwelling mammals (open symbols), the correlation is robust (West 1985). Generalized species with large heads (human, cow, elephant) and/or cochlear specializations (filled squares; horseshoe bat, dolphin, kangaroo rat, and gerbil) contribute most to the scatter in the data. Note that the mole rat fits the correlation for small, ground-dwelling mammals despite having a specialized cochlea.

Figure 5.12 shows the relationships between basilar membrane "stiffness" (ratio of thickness to width) and the high- (top panel) and low- (bottom panel) frequency limit of hearing. Stiffness of the basilar membrane at the base seems to be a good predictor of the upper frequency hearing limit, whether or not an animal is specialized (filled squares) or not (open circles). In contrast, the correlation between stiffness at the apex and the low-frequency hearing limit is poor, with outliers that are both specialized (horseshoe bat) and unspecialized (mouse). In general, the stiffer the basilar membrane is at the base of the cochlea, the higher the upper frequency limit of hearing is.

Both Masterton, Heffner, and Ravizza (1969) and Heffner and Heffner (1992) have provided evidence that the upper limit of hearing (at 60 dB SPL) in a wide range of mammals is correlated with effective head size (the time it takes for sound to travel from one ear to the other). As head size (and interaural distance) decreases, the sound frequencies required to generate

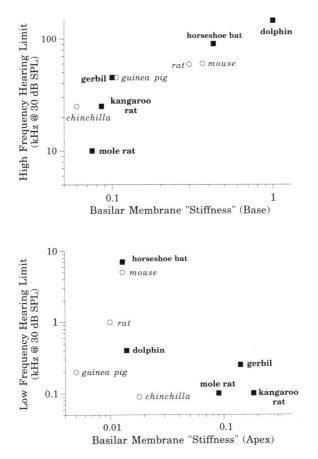

FIGURE 5.12. The upper panel illustrates the relationship between basilar membrane stiffness (thickness/width) at the base of the cochlea and high-frequency hearing limit for 10 mammals with specialized and generalized cochleae. The lower panel plots basilar membrane stiffness at the apex of the cochlea vs. the low-frequency limit of hearing for the same mammals.

detectable interaural intensity differences increases. The implication is that expansion of the high-frequency auditory range observed in many small mammals is an evolutionary adaptation for preserving sound localization ability.

6. Summary

This chapter has presented an overview of what we understand about the comparative structure of the cochlea of mammals. While there is a good body of data on the structure of the ear in several species, there is a surprising lack of data when considering the taxonomic breadth of extant mammals. Thus, there only are data for a few species, primarily those that have been used most as animal models in laboratory research and those showing cochlear specializations. Unfortunately, systematic and quantitative descriptions of cochlear morphology are not available for several widely used species having generalized cochleae (e.g., cat, guinea pig).

Clearly, it will be important to complete the database for the most commonly studied laboratory animals. In addition, however, it would be useful to obtain data on a number of other select species, such as additional data on more 'primitive' mammals. For example, there are only very limited data on the opossum, platypus, and echidna, species that may provide special insights into the evolution of the mammalian ear.

Several groups of specialized mammals have been of greatest interest to comparative anatomists. In particular, there is a wealth of data on hearing and the structure of the ear in various species of bats. There also are data on several other specialized species such as the mole rat and gerbil. However, a complete picture of the hearing capabilities in several of these species, such as the mole rat, is not available but is necessary in order to have a better understanding of the relationship between the structure and function of the inner ear. For example, while it is known that the mole rat cochlea is highly specialized in a very low-frequency region, there are no psychophysical data to tell whether and to what extent this species makes any special use of this frequency range and cochlear region in hearing.

Frequency-place maps for mammalian cochleae provide a way to organize thinking about both generalized and specialized structure-function relationships. The general cochlear map function of Greenwood (1961, 1962, 1990) has been quite useful in helping to discriminate between species that have generalized cochlear structures and functions and those that deviate qualitatively from this plan (the cochlear specialists). The specialists that have been most clearly identified to date (the horseshoe bat and mole rat) have abrupt transitions and steep slopes in their frequency-place maps that are closely associated with very shallow gradients of stiffness within particular cochlear regions. These specializations may or may not be clearly associated with the unusual thickening of the basilar membrane. The function of such specializations appears to be an enhanced frequency selectivity that results from a relatively large spatial separation between excitation-pattern maxima for closely spaced frequencies. The specializations for hearing, defined behaviorally, that arise from these morphological features are relatively well understood for some specialized bats, but remain to be further explored in the mole rat and other species (e.g., gerbils).

Finally, more precise structure-function relationships within the mammalian cochleae require a greater understanding of the interactions between macromechanical and micromechanical events occurring within the cochlear duct. In addition, considerably more experimental evidence is needed on the influence of specific accessory structures (such as the tectorial membrane) on the response properties of the cochlear partition.

Clearly, a better understanding of comparative issues in the mammalian inner ear depends upon a better understanding of the general micro- and macromechanical functions of the ear. At the same time, we would argue that we will gain a far better understanding of the function of the mammalian (and vertebrate!) ear when we have a broader appreciation of comparative structure and function of the ear in a wider range of mammalian species.

Acknowledgments. S. M. Echteler was supported by NIH Grant DC00493 during the preparation of this chapter.

References

Ashmore JF (1987) A fast motile response in guinea pig outer hair cells: the cellular basis of the cochlear amplifier. J Physiol 388:323–347.

Assad JA, Shepherd GMG, Corey DP (1991) Tip-link integrity and mechanical transduction in vertebrate hair cells. Neuron 7:985–994.

Bast TH, Anson BJ (1949) The Temporal Bone and the Ear. Springfield, IL: Charles C. Thomas.

Békésy G von (1952) DC resting potentials inside the cochlear partition. J Acoust Soc Am 24:72–76.

Békésy G von (1960) Experiments in Hearing. New York: McGraw-Hill Book Co.

Bloom W, Fawcett DW (1975). A Textbook of Histology, 10th Ed. Philadelphia: W. B. Saunders Co.

Bohne BA, Carr CD (1979) Location of structurally similar areas in chinchilla cochleas of different lengths. J Acoust Soc Am 66:411–414.

Bosher SK, Warren RL (1968) Observations on the electrochemistry of the cochlear endolymph of the rat. Proc R Soc London B 171:227–247.

Brownell WE, Bader CR, Bertran D, de Ribaupierre Y (1985). Evoked mechanical responses of isolated cochlear hair cells. Science 227:194–196.

Brundin L, Flock Å, Canlon B (1989). Sound-induced motility of isolated cochlear outer hair cells is frequency specific. Nature 342:814–816.

Brundin L, Flock Å, Khanna SM, Ulfendahl M (1992). The tuned displacement response of the hearing organ is generated by the outer hair cells. Neurosci 49:607–616.

Bruns V (1976) Peripheral auditory tuning for fine frequency analysis by the CF-FM bat, *Rhinolophus ferrumequinum*. I. Mechanical specializations of the cochlea. J Comp Physiol 106:77–86.

Bruns V (1979) Functional anatomy as an approach to frequency analysis in the mammalian cochlea. Verh Dtsch Zool Ges 1979:141–154.

Bruns V (1980) Basilar membrane and its anchoring system in the cochlea of the greater horseshoe bat. Anat Embryol 161:29–50.

Bruns V, Goldbach M (1980). Hair cells and tectorial membrane in the cochlea of the greater horseshoe bat. Anat Embryol 161:51–63.

Bruns V, Müller M, Hofer W, Heth G, Nevo E (1988) Inner ear structure and electrophysiological audiograms of the subterranean mole rat, *Spalax ehrenbergi*. Hear Res 33:1–10.

Bruns V, Burda H, Ryan MJ (1989) Ear morphology of the frog-eating bat (*Trachops cirrhosus*, Family: Phyllostomidae): apparent specializations for low-frequency hearing. J Morphol 199:103–118.

Burda H, Ballast L, Bruns V (1988) Cochlea in old world mice and rats (Muridae). J Morphol 198:269–285.

Canlon B, Brundin L, Flock Å (1988) Acoustic stimulation causes tonotopic alterations in the length of isolated outer hair cells from guinea pig hearing organ. Proc Natl Acad Sci USA 85:7033–7035.

Cody AR, Russell IJ (1987) The response of hair cells in the basal turn of the guinea pig cochlea to tones. J Physiol 383:551–569.

Dannhof BJ, Bruns V (1991) The organ of Corti in the bat *Hipposideros bicolor*. Hear Res 53:253–268.

Dallos P (1970) Low-frequency auditory characteristics: species dependence. J Acoust Soc Am 48:489–499.

Dallos P (1985a) Response characteristics of mammalian cochlear hair cells. J Neurosci 5:1591–1608.

Dallos P (1985b) The role of outer hair cells in cochlear function. In: Correia MJ, Perachico AA (eds) Contemporary Sensory Neurobiology. New York: Alan R. Liss, pp. 207–230.

Dallos P (1988a) The cochlea and auditory nerve. In: Edelman GM, Gall WE, Cowan WM (eds) Auditory Function. New York: John Wiley and Sons, pp. 153–188.

Dallos P (1988b) Cochlear neurobiology: revolutionary developments. ASHA June/July:50–56.

Dallos P (1992) The active cochlea. J Neurosci 12:4575–4585.

Dallos P, Evans BN, Hallworth R (1991) Nature of the motor element in the electrokinetic shape changes of cochlear outer hair cells. Nature 350:155–157.

DeRosier DJ, Tilney LG, Engelman E (1980) Actin in the inner ear: the remarkable structure of the stereocilium. Nature 287:291–296.

Ehret G, Frankenreiter M (1977) Quantitative analysis of cochlear structures in the house mouse in relation to mechanisms of acoustical information processing. J Comp Physiol A 122:65–85.

Engström H, Ades HW, Hawkins JE (1962) Structure and functions of the sensory hairs of the inner ear. J Acoust Soc Am 34:1356–1363.

Fay RR (1988) Hearing in Vertebrates: A Psychophysics Databook. Winnetka, IL: Hill-Fay Associates.

Fay RR (1992) Structure and function in sound discrimination among vertebrates. In: Webster DB, Fay RR, Popper AN (eds) The Evolutionary Biology of Hearing. New York: Springer-Verlag, pp. 229–263.

Fernández C (1952) Dimensions of the cochlea (guinea pig). J Acoust Soc Am 24:519–523.

Flock Å, Strelioff D (1984) Studies on hair cells from isolated coils of the guinea pig cochlea. Hear Res 15:11–18.

Goldberg JM, Fernández C (1984) The vestibular system. In: Darien-Smith I (ed) Handbook of Physiology. Section 1: The Nervous System. Volume III. Sensory Processes, Part 2. Bethesda, MD: American Physiological Society, pp. 977–1022.

Greenwood DD (1961) Critical bandwidth and the frequency coordinates of the basilar membrane. J Acoust Soc Am 33:1344–1356.

Greenwood DD (1962) Approximate calculation of the dimensions of traveling wave envelopes in four species. J Acoust Soc Am 34:1364–1369.

Greenwood DD (1990) A cochlear frequency-position function for several species—29 years later. J Acoust Soc Am 87:2592–2605.

Heffner RS, Heffner HE (1992) Hearing and sound localization in blind mole rats (*Spalax ehrenbergi*). Hear Res 62:206–216.

Henson MM, Henson OW (1991) Specializations for sharp tuning in the mustache bat: the tectorial membrane and spiral limbus. Hear Res 56:122–132.

Hoshino T (1976) Attachment of the inner sensory cell hairs to the tectorial membrane: A scanning electron microscopic study. Otorhinolaryngol 38:11–18.

Hudspeth AJ, Jacobs R (1979) Stereocilia mediate transduction in vertebrate hair cells. Proc Natl Acad Sci USA 76:1506–1509.

Iurato S (1962) Functional implications of the nature and submicroscopic structure of the tectorial and basilar membranes. J Acoust Soc Am 34:1386–1395.

Johnstone BM, Sellick PM (1972) The peripheral auditory apparatus. Q Rev Biophys 5:1–57.

Kachar B, Brownell WE, Altschuler R, Fex J (1986) Electrokinetic shape changes of cochlear outer hair cells. Nature 322:365–368.

Kiang NY-S, Watanabe T, Thomas EC, Clark LF (1965) Discharge Patterns of Single Fibers in the Cat's Auditory Nerve. Cambridge, MA: MIT Press.

Kiang NY-S, Liberman MC, Gage JS, Northrop CC, Dodds LW, Oliver ME (1984) Afferent innervation of the mammalian cochlea. In: Bolis L, Keynes RD, Maddrell SHP (eds) Comparative Physiology of Sensory Systems. Cambridge: Cambridge University Press, pp. 143–161.

Kimura RS (1966) Hairs of the cochlear sensory cells and their attachment to the tectorial membrane. Acta Otolaryngol 61:55–72.

Kronester-Frei A (1979) The effect of changes in endolymphatic ion concentrations on the tectorial membrane. Hear Res 1:81–94.

Lay D (1972) The anatomy, physiology, functional significance, and evolution of specialized hearing organs of gerbilline rodents. J Morphol 138:41–120.

Liberman MC (1982) The cochlear frequency map for the cat: labeling auditory nerve fibers of known characteristic frequency. J Acoust Soc Am 72:1441–1449.

Lim DJ (1972) Fine morphology of the tectorial membrane: its relationship to the organ of Corti. Arch Otolaryngol 96:199–215.

Lim DJ (1980) Cochlear anatomy related to cochlear micromechanics. A review. J Acoust Soc Am 67:1686–1695.

Lim DJ (1986) Functional structure of the organ of Corti: a review. Hear Res 22:117–146.

Long GR (1977) Masked auditory thresholds from the bat, *Rhinolophus ferrumequinum*. J Comp Physiol A 116:247–255.

Long GR (1980) Some psychophysical measurements of frequency processing in the greater horseshoe bat. In: van der Brink G, Bilsen FH (eds) Psychophysical, Physiological, and Behavioral Studies in Hearing. Delft, The Netherlands: Delft University Press, pp. 132–135.

Lowenstein O, Roberts TDM (1949) The equilibrium function of the otolith organs of the thornback ray (*Raja clavata*). J Physiol (London) 110:392–415.

Lowenstein O, Sand A (1940) The individual and integrated activity of the semicircular canals of the elasmobranch labyrinth. J Physiol (London) 99:89–101.

Masterton B, Heffner HE, Ravizza R (1969) The evolution of human hearing. J Acoust Soc Am 45:966–985.

Müller M (1991) Frequency representation in the rat cochlea. Hear Res 51:247–254.

Müller M, Ott H, Bruns V (1991) Frequency representation and spiral ganglion cell density in the cochlea of the gerbil *Pachyuromys duprasi*. Hear Res 56:191–196.

Müller M, Laube B, Burda H, Bruns V (1992) Structure and function of the cochlea of the african mole rat (*Cryptomys hottentotus*): evidence for a low frequency acoustic fovea. J Comp Physiol A 171:469–476.

Müller M, Wess FP, Bruns V (1993) Cochlear place-frequency map in the marsupial *Monodelphis domestica*. Hear Res 67:198–202.

Nadol J (1988) Comparative anatomy of the cochlea and auditory nerve in mammals. Hear Res 34:253–266.

Offner FF, Dallos P, Cheatham MA (1987) Positive endocochlear potential: mechanism of production by marginal cells of stria vascularis. Hear Res 29:117–124.

Pickles JO, Comis SD, Osborne MP (1984) Cross-links between stereocilia in the guinea pig organ of Corti and their possible relation to sensory transduction. Hear Res 15:103–112.

Plassmann W, Brändle K (1992) Functional model of the auditory system in mammals and its evolutionary implications. In: Webster DB, Fay RR, Popper AN (eds) Evolutionary Biology of Hearing. New York: Springer-Verlag, pp. 637–653.

Plassmann W, Peetz W, Schmidt M (1987) The cochlea of gerbilline rodents. Brain Behav Evol 30:82–101.

Pollak GD (1992) Adaptations of basic structures and mechanisms in the cochlea and central auditory pathway of the mustache bat. In: Webster DB, Fay RR, Popper AN (eds) Evolutionary Biology of Hearing. New York: Springer-Verlag pp.751–778.

Pujol R, Lenoir M, Ladrech S, Tribillac F, Rebillard G (1992) Correlation between the length of outer hair cells and the frequency coding of the cochlea. In: Cazals Y, Demany K, Horner K (eds) Auditory, Physiology and Perception, Advances in the Biological Sciences, Volume 83. New York: Pergamon Press, pp. 45–52.

Raphael Y, Lenoir M, Wroblewski R, Pujol R (1991) The sensory epithelium and its innervation in the mole rat cochlea. J Comp Neurol 314:367–382.

Retzius G (1884) Das Gehörorgan der Wirbeltiere. II. Das Gehörorgan der Reptilen, der Vögel und Säugetiere. Stockholm: Samson and Wallin.

Rhode WS (1984) Cochlear mechanics. Ann Rev Physiol 46:231–246.

Rose J, Brugge J, Anderson D, Hind J (1967) Phase-locked response to low frequency tones in single auditory, nerve fibers of the squirrel monkey. J Neurophysiol 30: 769–793.

Rosowski JJ (1992) Hearing in transitional mammals: predictions from the middle ear anatomy and hearing capabilities of extant mammals. In: Webster DB, Fay RR, Popper AN (eds) Evolutionary Biology of Hearing: Neuroanatomy. New York: Springer-Verlag, pp. 615–631.

Ryan A (1976) Hearing sensitivity of the mongolian gerbil, *Meriones unguiculatus*. J Acoust Soc Am 59:1222–1226.

Ryugo DK (1992) The auditory nerve: peripheral innervation, cell body morphology, and central projections. In: Webster DB, Popper AN, Fay RR (eds) The Mammalian Auditory Pathway: Neuroanatomy. New York: Springer-Verlag, pp. 23–65.

Saunders JC, Garfinkle TJ (1983) Peripheral anatomy and physiology I. In: Willot JF (ed) The Auditory Psychobiology of the Mouse. Springfield, IL: Charles C. Thomas, pp. 131–168.

Smith CA (1981) Recent advances in structural correlates of auditory receptors. In: Autrum H, Ottosen D, Perl E, Schmidt RF (eds) Progress in Sensory Physiology. Berlin: Springer-Verlag, pp. 135–187.

Smith CA, Tanaka K (1975) Some aspects of the structure of the vestibular apparatus. In: Naunton RF (ed) The Vestibular System. New York: Academic Press, pp. 3–20.

Spoendlin H (1966) The organization of the cochlear receptor. Adv Otorhinolaryngol 13:1–227.

Thorne PR, Carlisle L, Zajic G, Schacht J, Altschuler RA (1987) Differences in the distribution of F-actin in outer hair cells along the organ of Corti. Hear Res 30: 253–266.

Tilney LG, DeRosier DJ, Mulroy MJ (1980) The organization of actin filaments in the stereocilia of cochlear hair cells. J Cell Biol 86:244–259.

Vater M (1988) Cochlear physiology and anatomy in bats. In: Nachtigall PE, Moore PWB (eds) Animal Sonar, Processes and Performance. NATO ASI Series, Series A: Life Sciences, Volume 156. London: Plenum Press, pp. 225–241.

Vater M, Lenoir M (1992) Ultrastructure of the horseshoe bat's organ of Corti. I. Scanning electron microscopy. J Comp Neurol 318:367–379.

Vater M, Feng AS, Betz M (1985) An HRP-study of the frequency-place map of the horseshoe bat cochlea: morphological correlates of the sharp tuning to a narrow frequency band. J Comp Physiol A 157:671–686.

Voldrich L (1978) Mechanical properties of the basilar membrane. Acta Otolaryngol (Stockholm) 86:331–335.

Wada T (1923) Anatomical and physiological studies on the growth of the inner ear in the albino rat. Am Anat Mem 10:1–74.

Warr WB (1992) Organization of olivocochlear efferent systems in mammals. In: Webster DB, Popper AN, Fay RR (eds) The Mammalian Auditory Pathway: Neuroanatomy. New York: Springer-Verlag, pp. 410–448.

Webster DB, Webster M (1977) Auditory systems of Heteromyidae: cochlear diversity. J Morphol 152:153–170.

Webster DB, Webster M (1980) Morphological adaptations of the ear in the rodent family Heteromyidae. Am Zool 20:247–254.

West CD (1985) The relationship of the spiral turns of the cochlea and the length of the basilar membrane to the range of audible frequencies in ground dwelling mammals. J Acoust Soc Am 77:1091–1101.

Wever EG, Lawrence M (1954) Physiological Acoustics. Princeton, NJ: Princeton University Press.

Wever EG, McCormick JG, Palin J, Ridgway SH (1971a) The cochlea of the dolphin, *Tursiops truncatus*: general morphology. Proc Natl Acad Sci USA 68:2381–2385.

Wever EG, McCormick JG, Palin J, Ridgway SH (1971b) The cochlea of the dolphin, *Tursiops truncatus*: the basilar membrane. Proc Natl Acad Sci USA 68:2708–2711.

Wever EG, McCormick JG, Palin J, Ridgway SH (1971c) The cochlea of the dolphin, *Tursiops truncatus*: hair cells and ganglion cells. Proc Natl Acad Sci 68:2908–2912.

Wright A (1984) Dimensions of the stereocilia in man and guinea pig. Hear Res 22:89–98.

Young ED, Sachs MB (1979) Representation of steady-state vowels in the temporal aspects of the discharge patterns of populations of auditory nerve fibers. J Acoust Soc Am 66:1381–1404.

Zwislocki JJ (1986) Analysis of cochlear mechanics. Hear Res 22:155–169.

Zwislocki JJ, Kletsky EJ (1979) Tectorial membrane: a possible effect on frequency analysis in the cochlea. Science 204:639–641.

6
Outer and Middle Ears

JOHN J. ROSOWSKI

1. Introduction

The function of the mammalian external and middle ears (at least in terrestrial mammals) appears qualitatively similar. The external ear collects sound power and couples the collected power to the middle ear, and the middle ear transmits the power to the inner ear via motion of the tympanic membrane and ossicles. However, there are large differences in the scale and form of mammalian middle and external ears (Fig. 6.1), e.g., the African elephant (*Loxodonta africana*) has an external ear flap or pinna with an area of about 10^6 (mm)^2 and a tympanic membrane area of almost 10^3 (mm)^2, whereas the dwarf shrew (*Suncus etruscus*) has a pinna flap of only 10 (mm)^2 and a tympanic membrane area of only 1 (mm)^2 (Fleischer 1973; Heffner, Heffner, and Stichman 1982). There are also differences in the orientation and relative size of the ossicles (Fig. 6.1). In *Loxodonta*, the linear dimensions of the malleus are about twice those of the incus and the long arm of the malleus (the manubrium) is nearly vertical (perpendicular to the horizontal plane). In *Suncus*, the linear dimensions of the malleus are three to four times those of the incus and the manubrium of the malleus runs nearly parallel to the horizontal plane.

Quantitative comparisons of the performance of the ears of different mammals show large distinctions. For example, the audiograms of mammalian species show great differences in the range of sound frequencies to which each species is most sensitive, with the result that there are interspecific variations in hearing sensitivity of many orders of magnitude (Fay 1988; Long, Chapter 2). In general, the largest terrestrial mammals are most sensitive to sound frequencies below 10 kHz, while the smallest mammals are most sensitive to sound frequencies greater than 10 kHz (Heffner and Masterton 1980; Rosowski and Graybeal 1991; Heffner and Heffner 1992). The physiological reasons for this size-related segregation of hearing capabilities in mammals are far from completely understood, although there are a few studies that demonstrate that the high-frequency limits of hearing in a few species are inversely correlated to the dimensions of the tympanic membrane and ossicles

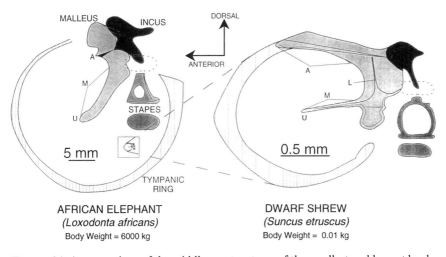

FIGURE 6.1. A comparison of the middle ear structures of the smallest and largest land mammals—medial views of the ring portion of the tympanic bone and ossicles of the African elephant and the dwarf shrew. The entire elephant tympanic ring is not illustrated. The malleus and incus are illustrated as they would be seen looking out from the oval window. The normal position of the stapes footplate is illustrated by a dotted ellipse, but the stapes has been disconnected from the incus in order to see the other ossicles. The disconnected stapes is shown in two views, including a medial view of the footplate and a superior view showing the stapes head and crura. Differences in shading are used to identify the three ossicles and the tympanic ring. The large drawings of the two ears are done at separate scales (scale bars are included). Inset into the tympanic ring of the elephant is a second illustration of the shrew middle ear drawn to the same scale as the elephant middle ear. Labels: A, the anterior process of the malleus; U, the umbo of the malleus; M, the manubrium of the malleus; L, the bony lamina which connects the manubrium and the anterior process of the malleus in the shrew ear. (After Fleischer 1973 with permission.)

(Khanna and Tonndorf 1978; Rosowski and Graybeal 1991; Rosowski 1992). There is also correlational evidence that the frequency-selective function of the external and middle ears has a large influence on the shape of the audiogram (Khanna and Tonndorf 1969; Dallos 1973; Zwislocki 1975; Rosowski, Carney, Lynch, and Peake 1986; Rosowski 1991), but these studies do not elucidate the structural causes for the observed correlations. In general, auditory science has not yet defined a working set of rules to describe the relationships between external and middle ear structures and functions, and our empirical knowledge of middle and external ear function comes primarily from a few species with ears of similar structure. The result is that precise relationships between structural features of the external and middle ears and auditory function are not well understood.

The primary goal of this chapter is to develop a framework for comparing external and middle ear function across mammalian species. In developing this framework, I will review measurements of middle and external ear function in mammalian species and discuss current theories relating structure and function. I will discuss two of the best understood middle ear structural features in some detail: the volume of the middle ear air spaces and the stiffness of the tympanic membrane, ossicles, and ligaments. The chapter concentrates primarily on the external and middle ears of terrestrial mammals, but the ears of marine mammals are also discussed. One topic that will not be discussed is the evolutionary development of the mammalian middle ear; although the accepted views on this subject have changed greatly in the last twenty years, they are the point of several recent reviews (Kermack and Musset 1983; Allin 1986; Rosowski and Graybeal 1991; Allin and Hopson 1992; Rosowski 1992) and need not be elaborated here.

In summarizing a vast literature, I had, on occasion, to select from several similar measurements made in different laboratories. I tried to select the most complete measurement set, but I also had to exercise some judgment in extracting specific features from the published measurements. The details of such extraction procedures are described in the table and figure captions.

2. Theories of External and Middle Ear Functions

2.1 Overview of Anatomy and Function

The detailed anatomy of the mammalian external and middle ears has been described by many authors (Doran 1879; Cockerell, Miller, and Printz 1914; van der Klaauw 1931; Segall 1943, 1969, 1973; Wever and Lawrence 1954; Kobrak 1959; Kirikae 1960; Henson 1961, 1970, 1974; Hinchcliffe and Pye 1969; Fleischer 1973, 1978; Shaw 1974a; Pye and Hinchcliffe 1976; Stinson and Khanna 1989), but a brief overview is appropriate (Fig. 6.2).

The external ear is usually broken into three major subdivisions: (1) the cartilaginous and sometimes mobile pinna flange that protrudes from the head, (2) the funnel-like concha that connects the opening of the pinna to the narrower ear canal, and (3) the tubelike ear canal. The air space enclosed by these structures is sometimes called the auditory tube. The size and complexity of the three external ear subdivisions vary greatly among species. Examples of structural complexities include: the many convexities and concavities of the human pinna and concha (Teranishi and Shaw 1968; Shaw 1974a; Searle et al. 1975); the numerous constrictions and wall projections within the auditory tube in cats (Wiener, Pfeiffer, and Backus 1966; Rosowski, Carney, and Peake 1988); the prominence and mobility of the tragus in some chiropteran species (Pye and Hinchcliffe 1976; Lawrence and Simmons 1982; Guppy and Coles 1988); and the oblique termination of the canal by the tympanic membrane in humans (Stinson 1985, 1986), cats (Khanna and Stinson 1985), gerbils (Ravicz, Rosowski, and Voigt 1992) and most other mammals (DiMaio and Tonndorf 1978).

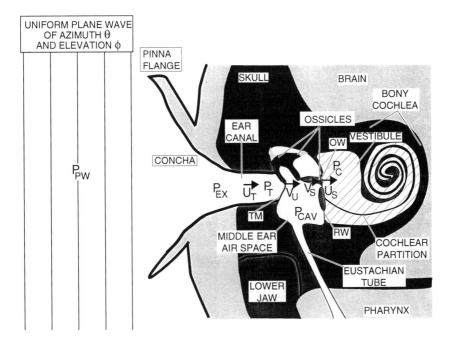

FIGURE 6.2. A schematic representation of the auditory periphery of a typical terrestrial mammal. Included in the schematic are the external ear (the pinna flange, concha, and ear canal), middle ear (the tympanic membrane, ossicles, middle ear air space, and Eustachian tube), and inner ear (the vestibule, cochlear windows, and cochlear partition). TM, OW, and RW refer to the tympanic membrane, oval window, and round window, respectively. The middle ear muscles (not illustrated) attach to the malleus (the tensor tympani muscle) and to the stapes (the stapedius muscle). Various mechanical and acoustic variables are also labeled. The illustrated sound pressures are those of the free-field uniform plane wave, $\mathbf{P}_{PW}$, the average pressure at the entrance to the ear canal, $\mathbf{P}_{EX}$, at the tympanic membrane, $\mathbf{P}_T$, in the middle ear air space, $\mathbf{P}_{CAV}$, and in the cochlear vestibule, $\mathbf{P}_C$. The volume velocities of the tympanic membrane, $\mathbf{U}_T$, and stapes footplate, $\mathbf{U}_S$, are also illustrated, as are the mechanical velocities of the umbo, $\mathbf{V}_U$ (the tip of the malleus handle at the center of the TM) and the stapes head, $\mathbf{V}_S$. Much of the discussion and analyses described in this chapter will assume the sinusoidal steady state in which any signal can be described in terms of the magnitude and "phase" angle of the response to a sinusoidal drive. The acoustic and mechanical variables are implicitly dependent on frequency, and each bold variable represents a complex quantity (containing real and imaginary components) that can be described in terms of a magnitude, $|\mathbf{P}|$, and an angle, $\angle\,\mathbf{P}$. The time (t) wave-form of the pressure and motion variables can be reconstructed as $p(t) = |\mathbf{P}| \cos(2\pi ft + \angle\,\mathbf{P})$, where f is the frequency of the sound stimulus.

The middle ears of terrestrial (and most marine) mammals contain several basic elements including: (1) a specialized tympanic membrane (TM) for the reception of sound; (2) an ossicular chain made up of three bony ossicles (in some mammals—notably the chinchilla *Chinchilla laniger*, guinea pig *Cavia procellus*, and house mouse *Mus musculus*—the incus and malleus are so tightly bound together that they appear fused) coupled and supported by several ligaments (Kobrak 1959; Kirikae 1960; Henson 1974); (3) an air-filled middle ear cavity; (4) a Eustachian tube to maintain aeration of the cavity; and (5) middle ear muscles that tense the tympanic membrane and the ossicular ligaments causing alterations in sound transmission through the ear (Borg 1968; Møller 1974). The size and shape of the tympanic membrane and ossicles differ greatly among species (Doran 1879; Cockerell, Miller, and Printz 1914; Henson 1961, 1974; Fleischer 1973, 1978; Pye and Hinchcliffe 1976; Hunt and Korth 1980; Kohllöffel 1984; Rosowski and Graybeal 1991), as do the size and structure of the middle ear air space (van der Klaauw 1931; Keen and Grobbelaar 1941; Legouix and Wisner 1955; Werner 1960; Webster 1965; Guinan and Peake 1967; Lay 1972; Hunt 1974; Webster and Webster 1975; Novacek 1977; Molvær, Vallersnes, and Kringlebotn 1978) and the relative size of the middle ear muscles (Henson 1961, 1970, 1974). The effects of some of these differences will be discussed in Section 4.1.

The inner ear acts as the load on the middle and external ears, and constrains the motion of the middle ear (Mundie 1963; Møller 1965; Lynch 1981). The inner ear structures that contribute to the load include the annular ligament (that holds the stapes footplate in the oval window), the vestibule (the fluid space just medial to the oval window), the fluid-filled scalae, the cochlear partition, and the round window membrane (Lynch, Nedzelnitsky, and Peake 1982; Puria and Allen 1991).

One can view external and middle ear function as a cascade of interdependent acoustical and mechanical processes with outputs that act as inputs to subsequent stages. Figure 6.3 is an illustration of such a view. The model has been constructed such that the input and output variables of each stage are restricted to measurable quantities. Another constraint is that the stages are all interdependent, i.e., the function of each stage is dependent on the function of later stages. Just as the middle ear is dependent on its cochlear load, the function of the external ear is dependent on the middle ear load, and changes in middle ear function can evoke changes in the measured "transfer function" of each external ear stage. (A transfer function is the ratio of an output to an input.)

The input to the ear is a uniform plane wave of sound pressure, P_{PW}, and direction of propagation relative to the ear as defined by the angles of azimuth θ and elevation ϕ. Interaction of the plane wave with the head, body, pinna, and concha results in a sound pressure, P_{EX}, and volume velocity, U_{EX}, at the ear canal entrance. (A volume velocity is a measure of the velocity of an acoustic medium integrated over an area. In the case of a uniform plane wave propagating through a tube, the volume velocity—with units of m^3/s—

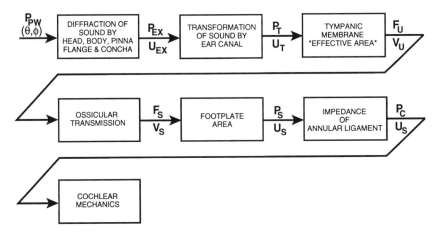

FIGURE 6.3. A "cascade" representation of the processes of the external and middle ears. Each block has associated input and output variables, and the processes within each box relate the inputs and outputs in a manner which is dependent on the load applied by "downstream" processes. The input to the ear is a plane wave of sound pressure, P_{PW}, and a propagation direction of azimuth θ and elevation ϕ. Many of the variables are described in the caption to Figure 2. Those not previously described are: U_{EX}, the volume velocity at the opening of the ear canal; F_U and F_S, the forces acting on the umbo and stapes; and P_S, the pressure applied by the stapes to the oval window. All of the output variables are similarly dependent on source azimuth θ and elevation ϕ, where these angles are defined by Shaw (1974a) such that θ and $\phi = 0$ describes the propagation direction from a sound source placed in front of the listener at the intersection of the midsagittal plane with the horizontal plane which contains the interaural axis.

can be calculated as the product of the average velocity of the fluid particles —with units of m/s—and the cross-sectional area of the tube.) P_{EX} and U_{EX} are dependent on direction (Shaw 1974a and b; Kuhn 1977), and this directional dependence remains unaltered throughout later signal transmission stages (Wiener and Ross 1946; Wiener, Pfeiffer, and Backus 1966; Hudde and Schröter 1980; Middlebrooks, Makous, and Green 1989).

The pressure and volume velocity at the ear canal entrance are transformed by the ear canal and the middle ear load into a pressure and volume velocity at the tympanic membrane, P_T and U_T. The tympanic membrane and the rest of the middle ear transform P_T and U_T into a force on and motion of the manubrium, noted in the schematic as a force and mechanical velocity at the umbo, F_U and V_U. (The umbo is the tip of the manubrium embedded near the center of the tympanic membrane.) The transformation from acoustic variables (pressure and volume velocity) to mechanical variables (force and velocity) can be approximated by the action of a piston of "effective" area, A_{ET}, such that $F_U = P_T \times A_{ET}$ and $V_U = U_T/A_{ET}$. As seen in Section 4.2.3 the effective areas defined by measurements of V_U and U_T are complex (have both magnitude and angle) and depend on sound frequency (Shera and Zwieg 1991; Lynch, Peake, and Rosowski 1994). Another complication is that the measured umbo velocity represents the sum of several mechanical processes including: (1) rotation about the ossicular axis (Dahmann 1929; Wever and Lawrence 1954; Kobrak 1959; Guinan and Peake 1967); (2) inward and outward translation of the tympanic membrane–malleus complex (Gyo, Aritomo, and Goode 1987; Decraemer, Khanna, and Funnell 1991; Donahue, Rosowski, and Peake 1991); and (3) possible bending of the manubrium (Funnell, Decraemer, and Khanna 1992).

The force, F_U, and velocity, V_U, applied to the umbo act on the rest of the middle ear and the cochlear load producing a force, F_S, and velocity, V_S, at the stapes footplate. This force and velocity are integrated over the effective area of the stapes footplate, A_{ES}, to produce a pressure acting on the cochlear window, P_S, and a volume velocity of the stapes footplate, U_S. If the stapes moves as a piston, then A_{ES} approximates the anatomical area of the footplate, A_{FP}. Part of the pressure applied to the oval window is taken up in flexing the annular ligament that holds the stapes in the window, and this decrease in pressure from the stapes, P_S, to the vestibule, P_C, is the final step of middle ear sound transformation. The sound pressure, P_C, and volume velocity, U_S, within the vestibule produce the inner ear fluid motions associated with motions of the cochlear partition.

While others have used mechanical velocity to quantify the motion of air in the ear canal or fluid in the vestibule (e.g., Allen 1986), I use the acoustic variable, volume velocity. The volume velocity represents motion integrated over an area, e.g., the area of the tympanic membrane or the stapes footplate. Direct measurements of volume velocity at the tympanic membrane, usually associated with measurements of acoustic impedance, readily account for variations in the motion across the surface of the membrane (Khanna and Tonndorf 1972). While the estimation of volume velocity can be straightforward, the determination of the average sound pressure at the tympanic membrane can be difficult at higher frequencies where the dimensions of the ear's structures begin to approximate or are larger than sound wave length (Khanna and Stinson 1985). In humans and cats, it is possible to specify precisely the sound pressure acting at the tympanic membrane only for frequencies below 10 kHz. The limit would be higher in animals with smaller external ears and tympanic membranes.

2.2 The Concept of Impedance Matching and the Ideal Transformer Model

The commonly held view of external and middle ear function has not changed much during the last 125 years. Helmholtz (1868) first described the middle ear in terms of hydraulic and rotational levers that transform the acoustical impedances of air and cochlear fluid in order to improve the coupling of sound power between these two media. The coupled sound then produces motions of the cochlear fluid via the differential stimulation of the cochlear windows. The pioneering studies of Dahmann (1929), Tröger (1930), Waetzmann and Keibs (1936), Metz (1946), Wever and Lawrence (1954), and Békésy (1960) are some of the earliest empirical tests of this hypothesis. In their monograph *Physiological Acoustics*, Wever and Lawrence (1954) summarized much of the earlier work, and described an extensive series of empirical measurements that built on the transformer hypothesis, and defined what has come to be known as the "ideal transformer model" of middle ear function (Dallos 1973). In this ideal view, the large difference in the anatomical area of the tympanic membrane, A_{TM}, and stapes footplate, A_{FP}, together with the mechanical advantage produced by a rotation of the longer malleus, L_M, and shorter incus, L_I, about a fixed axis—the ossicular lever—results in a gain of sound pressure within the cochlear vestibule where

$$\frac{P_C}{P_T} = \frac{A_{TM}}{A_{FP}} \frac{L_M}{L_I} \gg 1. \tag{1}$$

The model also predicts a concomitant reduction in the volume velocity between the tympanic membrane and the stapes, $U_S/U_T = P_T/P_C$. Thus, the proposed ideal transformer not only couples sound power from the low impedance of air to the high impedance of the cochlear fluid, it also allows for easy predictions of middle ear function (see, e.g., Wever and Lawrence 1954; Dallos 1973). A common variation of the ideal transformer model further assumes that the area ratio and ossicular lever arms are optimized to overcome the air cochlea impedance mismatch and has led to the idea that auditory function can be estimated from anatomical measurements of tympanic membrane and footplate areas and malleus and incus lengths.

Several authors of varied disciplines have used this attractive simplification to address issues as diverse as inter specific differences in auditory sensitivity (Lay 1972; Peterson et al. 1974; Webster and Webster 1975) and the hearing ability of extinct mammals and reptiles (Allin 1975; Kermack and Musset 1983). Unfortunately, as Wever and Lawrence (1954, Chapter 17) and Dallos (1973) explained later in their monographs, the ideal transformer model is a simplification; it ignores constraints placed on sound conduction by the mechanical and acoustical effects of the external and middle ear structures themselves. Just as it is difficult to produce an ideal electrical transformer in which the inductance and resistance are zero and the capacitance infinite, the stiffness, mass, and viscous losses within the tympanic membranes, ossicles, ossicular joints and middle ear air spaces constrain the ear's transformer (Dallos 1973; Zwislocki 1975; Killion and Dallos 1979; Schubert 1980; Shaw and Stinson 1983). These constraints cannot be ignored.

2.3 An Alternate View of External and Middle Ear Function

One class of alternatives to the ideal transformer model of middle and external ear function are what might be called "peripheral filter" hypotheses. These hypotheses contend that the acoustical and mechanical constraints of the external and middle ears (that are ignored by the ideal transformer model) determine the *shape* of the audiogram (Waetzmann and Keibs 1936; Khanna and Tonndorf 1969; Dallos 1973; Zwislocki 1975; Shaw and Stinson 1983; Rosowski et al. 1986; Rosowski 1991). Peripheral filter hypotheses are of three basic types:

(1) Isostapes motion theories correlate auditory threshold at various frequencies with the free-field plane wave sound pressure, $\mathbf{P}_{PW}$, required to produce a constant motion of the stapes, usually velocity (Guinan and Peake 1967; Dallos 1973; Zwislocki 1975).

(2) Isocochlear pressure theories correlate threshold with the plane wave sound pressure required to produce a constant sound pressure within the cochlea (Lynch, Nedzelnitsky, and Peake 1982; Décory 1989).

(3) Isopower theories correlate threshold with the plane wave sound pressure required to produce a given sound power at the entrance to the cochlea (Waetzmann and Keibs 1936; Khanna and Tonndorf 1969; Rosowski 1991).

The constant velocity, pressure, and power theories are not easily distinguished because the cochlear input impedance is resistive over a broad frequency range; hence, sound pressure functions that maintain constant power at different frequencies also produce constant velocity and pressure. Indeed, all of these theories adequately predict the audiogram shape of a small number of mammalian species (Dallos 1973; Zwislocki 1975; Lynch, Nedzelnitsky, and Peake 1982; Rosowski et al. 1986; Rosowski 1991).

The isopower theories do have several advantages over the isovelocity or pressure theories (Rosowski et al. 1986):

(1) Although they are distinct from the *ideal* transformer model, isopower theories are conceptually linked to the broader notion of the transformation of sound energy from air to the cochlea. The purpose of a transformer is to improve the transfer of power between the source and load impedance. The power theories include this transformation but allow for significant frequency-dependent losses between the power available from the stimulus and the power that reaches the inner ear.

(2) The power theories also account for interspecific variations in the impedance of the inner ear. Neither the isovelocity nor isopressure theories totally describe the input to the inner ear by themselves; the cochlear pressure associated with a given stapes velocity depends on the impedance of the cochlea. Interspecies differences in cochlear impedance produce uncertainties in the relationship between stapes velocity and vestibular sound pressure and greatly complicate interspecies comparisons of one or the other. Power theories take impedance into account.

(3) Measurements of power transfer efficiency are a standard tool for comparing the performance of electrical, mechanical, or acoustical systems, and such a quantitative measure would be useful in the interspecific comparisons of ear function.

These advantages make estimates of power flow in the ear useful for both understanding external and middle ear functions and comparing function among species.

2.4 Power Flow Through the Auditory Periphery

Figure 6.4 is a schematic that illustrates the flow of sound power through the external and middle ears. Again the auditory periphery is broken into connected blocks with generally well-defined and measurable input and output variables. The input to the first (left most) block is a uniform plane wave of known sound pressure and direction, just as in the cascade description (Fig. 6.3). Subsequent stages of the schematic are defined in terms of an electric analog in which sound pressure is analogous to voltage and volume velocity is analogous to current (Onchi 1961; Zwislocki 1962; Fletcher and Thwaites 1979). The left-most block of Figure 6.4 performs the function of the first two stages of Figure 6.3 and describes the transformation from $\mathbf{P_{PW}}$ to $\mathbf{P_T}$ and $\mathbf{U_T}$.

The middle ear block performs the functions of the third through sixth

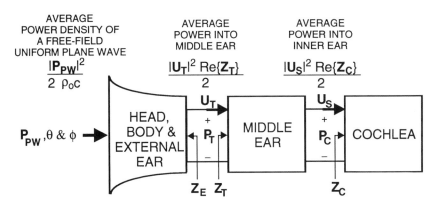

FIGURE 6.4. A schematic representation of the flow of average sound power from the stimulus to the cochlea. The left-most block, which accounts for the action of the head, body, and external ear, has a uniform plane wave of sound pressure, $\mathbf{P_{PW}}$, and direction, θ and ϕ, as its input, with outputs of volume velocity and pressure at the tympanic membrane, $\mathbf{U_T}$ and $\mathbf{P_T}$. The middle ear block is represented by a two-port network which can be defined in terms of its inputs, outputs (the volume velocity of the stapes, $\mathbf{U_S}$, and the sound pressure in the cochlear vestibule, $\mathbf{P_C}$), and load. The cochlea is a one-port defined completely by its input impedance, $\mathbf{Z_C}$. The radiation impedance looking out the external ear from the tympanic ring, $\mathbf{Z_E}$, and the middle ear input impedance, $\mathbf{Z_T}$, are also illustrated. The equations at the top define the average power or power density at the inputs to the three stages of the system. $\mathrm{Re}\{\mathbf{Z}\}$ is the real part of the impedance, where $\mathrm{Re}\{\mathbf{Z}\} = |\mathbf{Z}| \cos (\angle \mathbf{Z})$; ρ_0 is the density of air at normal atmospheric pressure; c is the speed of sound; and $\rho_0 c$ is the characteristic impedance of air. (After Rosowski 1991.)

stage of Figure 6.3, transforming $\mathbf{P}_T$ and $\mathbf{U}_T$ to $\mathbf{P}_C$ and $\mathbf{U}_S$, and is defined in terms of a two-port network (Desoer and Kuh 1969). A two-port, with two terminals at the input port and two terminals at the output, is a general circuit description tool. Any linear physical system with two interdependent inputs (e.g., force and velocity, voltage and current, or pressure and volume velocity) at one port and two interdependent outputs at another port can be defined in terms of such a network. The two-port representation also implies a dependence of the input and output variables on the load, i.e., alterations in the cochlea can have significant effects on the ratios of the input and output variables of the middle ear.

The load on the middle ear is the cochlea, modeled by a one-port (two-terminal) device that is completely defined by its impedance, $\mathbf{Z}_C = \mathbf{P}_C/\mathbf{U}_S$. The schematic also illustrates other impedances, including the middle ear input impedance, $\mathbf{Z}_T = \mathbf{P}_T/\mathbf{U}_T$, and the impedance looking out the ear tube from the tympanic membrane, $\mathbf{Z}_E$, the so-called external ear radiation impedance (Siebert 1973; Shaw 1976, 1988; Rosowski, Carney, and Peake 1988).

The practical significance of the power flow scheme is twofold: all the variables used to define the inputs and outputs of the different ear parts have been measured in several animals, and these measurements allow one to define and compare the power that enters each block (Rosowski et al. 1986). Calculations of power flow through the ear have been made for several mammalian species including humans, cats, chinchillas, and guinea pigs (Shaw and Stinson 1983; Rosowski et al. 1986; Rosowski 1991). The results of these calculations, which are discussed in Section 4.2.7, generally support the hypothesis that the filtering of acoustic power by the external and middle ears plays a large role in determining the shape of the audiogram.

2.5 Relationships Between Specific Structures and Auditory Function

Although the system diagram of Figure 6.3 and the power flow diagram of Figure 6.4 define frameworks for quantifying external and middle ear function, they do not associate specific structures with auditory performance. What is needed is to replace each of the boxes in Figures 6.3 and 6.4 with a set of equations or a network of elements that specifies the role of individual structures in the function of the box. However, a complete mathematical description of what is known about the structure and function of the external and middle ear is beyond the scope of this chapter. Those readers interested in such a description should look to other sources, notably the circuit descriptions of Møller (1961), Onchi (1961), Zwislocki (1962, 1963, 1975), Peake and Guinan (1967), Marquet, Van Kamp, and Creten (1973), Matthews (1980, 1983), Kringlebotn (1988), and Wada and Kobayashi (1990). More detailed associations between structure and function are also available (Funnell and Laszlo 1977; Rabbit and Holmes 1986; Funnell, Decraemer, and

Khanna 1987,1992; Puria and Allen 1991), but the mathematical description of middle ear structure and function is far from complete. More work needs to be done, at both the conceptual and data gathering levels. Indeed, the major deficiency limiting attempts to describe precise relationships between structure and function in the middle and external ears is a lack of detailed functional and structural measurements.

3. Measurements of External Ear Function and Structure

The external ear has two distinct acoustic roles: (1) the collection and delivery of sound power from the environment to the middle ear, and (2) the imposition of directional dependence on the power collection process, where this directionality provides the acoustic cues used in localizing sounds in space (Khanna and Tonndorf 1978; Blauert 1983; Kuhn 1987; Brown, Chapter 3). For most localized stimulus sources, the power collected by the ear is a function of source direction and the two functions of the external ear cannot be easily separated. There are, however, specialized nondirectional stimulus conditions, e.g., diffuse sound fields (Beranek 1949; Shaw 1988), that can be used to isolate the power collection performance of the ear (Shaw 1988). This section will discuss functional measurements that quantify external ear power collection and directionality and attempt to relate those functions to ear structure. Nonacoustic functions of the external ear will not be discussed, e.g., thermoregulation in elephants (Buss and Estes 1971) and rabbits (Hill, Christian, and Veghte 1980).

3.1 Comparisons of Size and Structure

As noted previously, the external ear varies greatly in size and shape among mammals. The structures of the external ears of three mammals are compared in Figure 6.5. The drawings illustrate some of the fine structure in external ear configuration including: (1) variations in the cross-sectional area of the ear canal of as much as a factor of two along the canal length; (2) bends in the auditory tube near the connection of the concha and canal; and (3) numerous invaginations and protuberances of the cartilaginous walls of the pinna flange and concha. Some of this fine structure provides cues for determining the elevation of high-frequency sound sources (Teranishi and Shaw 1968; Shaw 1974a,b, 1982; Searle et al. 1975; Butler and Belendiuk 1977; Musicant, Chan, and Hind 1990).

The location of the ear on the head and the ratio of head-to-ear size also vary among these three species. The external ears of humans are located on the side of the head and the pinna flange is much smaller than the head. In contrast, the Tammar wallaby (*Macropus eugenii*) has pinna flanges that are nearly the size of its head, and the pinnae are placed high on the head close to the superior aspect of the skull (Coles and Guppy 1986). The domestic cat

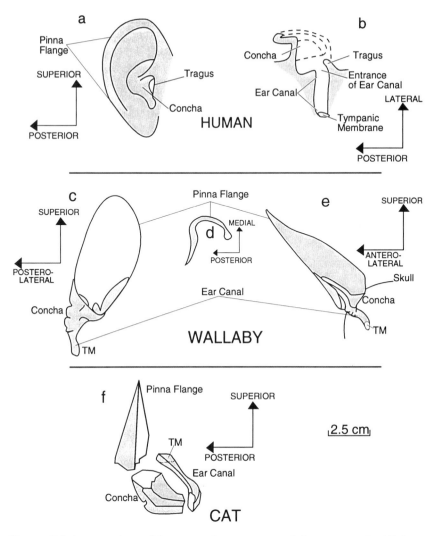

FIGURE 6.5. A comparison of the external ear anatomy of three mammals. All draw-ings are to the same scale. (a and b) The human external ear (after Shaw 1974a with permission): (a) a lateral view of the pinna flange, concha, and entrance of the ear canal (shaded white); and (b) a horizontal section of the ear at the level of the canal; dashed lines are used to illustrate some of the structures superior to the plane of section. (c, d, and e) The external ear of the Tammar wallaby *Macropus eugenii* (after Coles and Guppy 1986 with permission): (c) a view from the front and side; (d) a horizontal cross-section midway through the pinna; and (e) a near frontal view show-ing the normal tilt of the ear and canal and some of the complexities of the surface of the pinna. (f) A three-dimensional reconstruction of the external ear of a domestic cat *Felis catus* (after Rosowski, Carney, and Peake 1988), illustrating the pinna flange, funnel-shaped concha, and ear canal. Note the variation in the diameter of the canal with length and the notches and convolutions within the concha.

TABLE 6.1. Dimensions of the external ears of ten mammals described in the literature. The parts of the external ear are defined in Figure 5. The pinna flange is the visible pinna flap. The ear tube is the air space enclosed by the external ear. The area of the pinna opening is an estimate of the opening of the external ear. The concha is defined within the various references. The ear canal is the relatively straight and uniform tube between the tympanic membrane (TM) and the funnel-like concha.

Species	Author	Body weight (kg)	Height of pinna flange (mm)	Total length of ear tube (mm)	Area of pinna opening (mm²)	Area of concha opening (mm²)	Volume of ear canal (μl)	Cross-sectional area of ear canal (mm²)	Length of ear canal (mm)
Cat (Felis catus)	Wiener, Pfeiffer, and Backus 1966			20 (TM to tragus)					15
	Shaw 1974a		40				230	15	
	Rosowski, Carney, and Peake 1988	2.5		22.4–34.3	1250	230		19–40	9.5–18.9
Chinchilla (Chinchilla laniger)	Dear 1987	0.8		46	760			56	32
	Shaw 1974a							28	30
Ferret (Mustela putorious)	Carlile 1990b	0.3	40				840		10–12
Ghost bat (Macroderma gigas)	Guppy and Coles 1988	0.13		35	908		550	50	11
Gleaning bat (Nyctophilus gouldi)	Guppy and Coles 1988	0.01		15.5	227		31.5	12.6	2.5
Guinea pig (Cavia procellus)	Shaw 1974a	0.4				20	45	5	9
Human	Shaw 1974a	75	67	12		433	1000	44	22.5
Mongolian gerbil (Meriones unguiculatus)	Ravicz 1990	0.1			71			4.3	
Mouse (Mus musculus)	Saunders and Garfinkle 1983	0.05		6.25 (TM to tragus)					
Tammar wallaby (Macropus eugenii)	Coles and Guppy 1986	4	70	61	1960		800	50	16

also has large pinna flanges, but the base of the ear is placed on the side of the head and only the top two-thirds of the flange protrudes above the head. The pinna flanges of the cat and wallaby are highly mobile, unlike the fixed pinnae of humans.

There has been little quantitative work comparing the dimensions of the external ears of mammals. The most comprehensive comparison until this time has been that of Shaw (1974a). I have summarized Shaw's data and added some newer information (Table 6.1). Comparisons of the data in Table 6.1 are complicated by the intricate geometry of the ear tube as well as interlaboratory differences in the methodology used to define and measure structural features. Not surprisingly, the data suggest that smaller mammals tend to have smaller external ears. There are exceptions to this rule, e.g., the wallaby has a taller pinna flange and wider ear canal than the human. An adequate description of how the external ear dimensions scale with body size requires more inter- and intraspecific measurements of structure.

3.2 Measurements of Function

3.2.1 External Ear Gain: Magnitude and Directional Dependence

The most common measurement of external ear function has been the magnitude of the ratio of sound pressure at the tympanic membrane to the sound pressure of a plane wave stimulus $|\mathbf{P}_T/\mathbf{P}_{PW}|$ (e.g., Wiener and Ross 1946; von Bismark 1967; Drescher and Eldredge 1974). This ratio has been called the "external ear pressure gain." Measurements of this gain in nine different species are illustrated in Figure 6.6. In all of the species, the gain is close to 0 dB at the low frequencies and is maximum between 2 and 30 kHz. The maximum gain can be as large as 27 dB. The frequency of maximum gain is sometimes called the external ear resonance frequency. This frequency appears highest in the four smallest species in the comparison population (the two bats, the ferret, and the mouse), but there is only a loose inverse relationship between canal length (Table 6.1) and the maximum frequency. One confounding aspect of the data in Figure 6.6 is that the external ear gain is directional; the gain varies with source position, especially with changes in source azimuth (Wiener and Ross 1946; Wiener, Pfeiffer, and Backus 1966; Shaw 1974a). Details of source direction for each measurement are included in the caption. Most of the illustrated measurements were performed with a single source position, usually on the horizontal plane; however, the pictured external ear gains in the two bats and wallaby were measured at the acoustic axis (direction of largest pressure gain) that varied with frequency.

Some more recent measurements of external ear pressure gain demonstrate a large dependence on elevation of the sound source (Fig. 6.7) as well as the azimuth. The most prominent elevation-related changes in gain occur at frequencies above the external ear resonance frequency and result from elevation-dependent changes in the frequencies of gain minima (Shaw 1982; Musi-

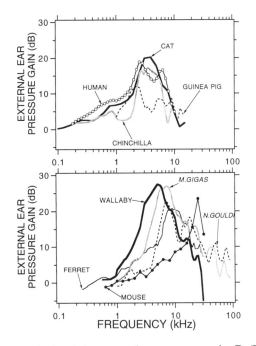

FIGURE 6.6. The magnitude of the external ear pressure gain, P_T/P_{PW}, measured in nine mammalian species. The measurements are made with different source directions. Cat (*Felis catus*): azimuth of 90° and elevation of 0° (Wiener, Pfeiffer, and Backus 1966). Chinchilla: 0° and 0° (von Bismark and Pfeiffer 1967). Guinea pig: 90° and 0° (Sinyor and Laszlo 1973). Human: 90° and 0° (Shaw 1974a). Ferret (*Mustela putorious*): 30° and 15° (Carlile 1990b). House mouse (*Mus musculus*): 90° and 0° (calculated from Saunders and Garfinkle 1983 by assuming a head shadow of 6 dB at frequencies above 20 kHz). Wallaby (*Macropus eugenii*): at the acoustical axis of each frequency (Coles and Guppy 1986). Two bats (*Macroderma gigas* and *Nyctophilus gouldi*): at the acoustical axis of each frequency (Guppy and Coles 1988). In general, interspecific variations are much larger than the variations within species.

cant, Chan, and Hind 1990). These minima seem to result from interference between reflections of the sound produced by different fine structural features within the pinna flange and concha (Teranishi and Shaw 1968; Shaw 1974a,b, 1982; Searle et al. 1975; Musicant, Chan, and Hind 1990).

3.2.2 Direct Measurements of Directionality

The directionality of the external ear has been assessed by the measurement of the effect of sound source azimuth and elevation on the ear's response to tones of different frequencies (Phillips et al. 1982; Calford and Pettigrew 1984; Coles and Guppy 1986; Carlile and Pettigrew 1987; Guppy and Coles 1988; Jen and Chen 1988; Carlile 1990a,b). These measurements define elliptical

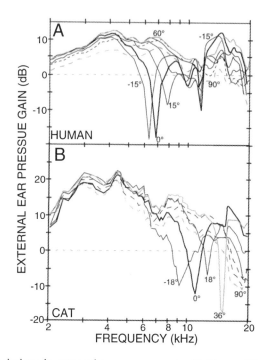

FIGURE 6.7. Variations in external ear pressure gain, P_T/P_{PW}, with source elevation. Top panel: human data from Shaw (1982). Bottom panel: cat data from Musicant, Chan, and Hind (1990). All measurements were made with an azimuth of 0° and elevations of between −20° and 90°. The unlabeled lines were measured at elevations between 0° and 90°. (After Musicant, Chan, and Hind 1990 with permission.)

isogain contours in azimuth and elevation; the centers of these contours are the acoustic axes. The location of the acoustic axis is frequency dependent (Calford and Pettigrew 1984), as are the widths and heights of the isogain contours. The isogain contours are broad and nonspecific for low-frequency tones but very narrow and well localized in space for high-frequency tones. A quantitative measurement of the directional specificity of these contours is the "acceptance angle," defined as the width or height of the contour that specifies a 3-dB reduction (a 50% decrease in power) from the gain at the acoustic axis (Coles and Guppy 1986). In the animals where it has been measured (cat, guinea pig, wallaby, and two bats), this angle can be as small as 30° to 40° at frequencies above 10 kHz and larger than half of space, >180°, at frequencies below 1 kHz (Calford and Pettigrew 1984; Coles and Guppy 1986).

The variation in acceptance angle with frequency can be approximated by calculations of the diffraction of sound passing through a circular aperture at the end of a tube (Calford and Pettigrew 1984; Coles and Guppy 1986). Figure 6.8 compares the measured acceptance angle for the wallaby ear with

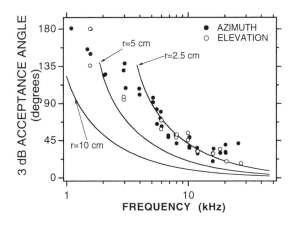

FIGURE 6.8. Spatial selectivity of the external ear of the Tammar wallaby measured in terms of the 3-dB (half-power) acceptance angle in the azimuthal (horizontal) and elevational (vertical) planes. The data are based on measurements of external ear pressure gain produced by a sound source placed at varied azimuths and elevations (Calford and Pettigrew 1984). The acceptance angle for azimuth is the horizontal width of the ear canal gain contour that demarcates a 3-dB reduction from the gain at the acoustic axis. The angle for elevation is the vertical height of the same contour. Each of the filled circles represents an azimuthal acceptance angle measured in one of four ears. The open circles are the elevational angles measured in two ears. The three lines are estimates of the directivity of sound diffraction through circular apertures of radii equal to 2.5, 5, and 10 cm. (After Coles and Guppy 1986.)

the angles predicted for aperture radii of 2.5 (the average radius of the wallaby's pinna opening; Coles and Guppy 1986), 5, and 10 cm. At frequencies between 4 and 20 kHz, the data are well fit by the anatomical radius, but, at lower frequencies, the data are better fit by larger apertures. The match between data and theory at frequencies above 4 kHz suggests that the directionality of the pinna at these frequencies is approximated by diffraction about the pinna opening, i.e., at high frequencies the structures surrounding the ear have little effect on the ear's directionality. The increased apparent aperture dimensions at lower frequencies may result from the influence of the non-uniform shape of the ear tube (Fletcher and Thwaites 1988) and contributions of the head and body to external ear directionality (Calford and Pettigrew 1984; Coles and Guppy 1986). It should be noted that, although high-frequency acceptance angles may be accurately predicted from estimates of diffraction about the ear opening, simple diffraction does not explain the elevation-related notches in the external ear gain observed in humans and cats (Shaw 1982; Musicant, Chan, and Hind 1990). As discussed in Section 3.2.1, these notches are thought to be produced by the interactions of reflections from different pinna flange and concha structures (Teranishi and Shaw 1968).

3.2.3 Radiation Impedance and Power Flow

As stated earlier in this section, one method for isolating the sound power collection properties of the external ear canal is to use a nondirectional stimulus like a diffuse sound field (Shaw 1976, 1988). Such stimuli are difficult to produce (Shaw 1976, 1979; Kuhn 1979), but the acoustic reciprocity principle suggests a simple method for measuring the diffuse field response (Shaw 1988). In a reciprocal system, for sound waves excited at any point, A, "the resulting velocity potential at a second point, B, is the same both in magnitude and phase, as it would have been at A had B been the source of the sound" (Rayleigh 1945). In the special case of a subject placed within a diffuse sound field, the relationship between the stimulus pressure, P_{DF}, and the volume velocity at the tympanic membrane of the subject is identical with the relationship between a sound source at the tympanic membrane of pressure, P_E, and the volume velocity, U_E, delivered to the diffuse field (Siebert 1970; Shaw 1988). Therefore, measurements of P_E and U_E produced by a sound source placed at the tympanic membrane looking out the ear can be used to

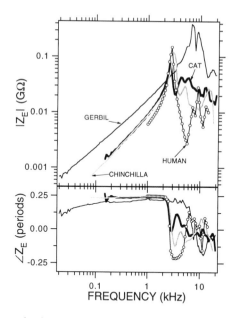

FIGURE 6.9. The magnitude $|Z_E|$ and angle $\angle Z_E$ of the radiation impedance of the external ear from four species. Domestic cat: mean of measurements from six ears (Rosowski, Carney, and Peake 1988). Chinchilla: mean from six ears (Dear 1987). Gerbi: mean from six ears (Ravicz 1990). Human: measurements from a model of the ear (Shaw 1976, 1979). The units of impedance magnitude are MKS acoustic gigaohms (1 GΩ = 10^9 Pa-s-m^{-3}). The angle of the impedance is scaled in periods, where 1 period equals 2π radians or 360°. The interspecific variation generally is much larger than the variation within species.

estimate the diffuse field response and the power collecting capabilities of the ear (Siebert 1970; Shaw 1976, 1979, 1988; Schröter and Poesselt 1986).

The ratio P_E/U_E is the radiation impedance looking out the external ear, Z_E (Rosowski, Carney, and Peake 1988), and this impedance has been measured in four mammalian species (Fig. 6.9). The four measurements have some similar features; at low frequencies the impedance is mass controlled (the magnitude of the impedance $|Z_E|$ is proportional to the frequency and its angle, $\angle Z_E$, approximates 0.25 periods), and there is a magnitude maximum at some midfrequency where the angle of the impedance is 0. The generally inverse relationship between impedance magnitude and ear size suggests that smaller ears have higher radiation impedances (Rosowski 1991).

The theoretical effect of changes in external ear dimensions on radiation impedance is illustrated in Figure 6.10. The radiation impedances in this figure were calculated from a circuit representation of a scaled acoustic model of the cat external ear. The model consisted of the combination of a uni-

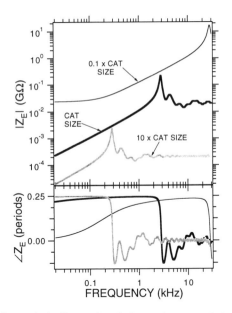

FIGURE 6.10. The theoretical effects of variations of ear canal size on the magnitude and angle of the radiation impedance of the external ear. The cat auditory tube model (Rosowski, Carney, and Peake 1988, Figure 11) consists of a coupled exponential horn and a uniform tube with dimensions similar to that of the cat's ear tube. The impedance of the model (which closely approximates the radiation impedance of the cat's external ear; Rosowski, Carney, and Peake 1988) was calculated with lossy uniform tube solutions (Egolf 1980; Zuercher, Carlson, and Killion 1988) for a concatenation of 100 tubes whose dimensions were chosen to approximate the model. The model calculations were repeated after scaling the linear dimensions of the cat-sized model by factors of 0.1 and 10.

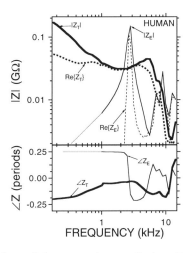

FIGURE 6.11. A comparison of the external ear radiation, Z_E, and middle ear input, Z_T, impedances estimated for the human ear. The radiation impedance is that of Shaw (1976, 1979). The middle ear input impedance is a combination of the measurements of Rabinowitz (1981) and Hudde (1983). The magnitudes of the real part of the impedances, where $Re\{Z\} = |Z| \cos (\angle Z)$, are also plotted.

form tube and an exponential horn (Rosowski, Carney, and Peake 1988, Fig. 6.11). The illustrated impedances include predictions for the cat-sized model and for models with linear dimensions scaled up and down by a factor of 10. Scaling the dimensions has five effects on the radiation impedance. (1) The magnitude of the low-frequency masslike impedance and (2) the frequency of the peak magnitude are approximately inversely proportional to the linear dimensions. (3) The peak $|Z_E|$ and (4) the resistive high-frequency impedance appear inversely proportional to the square of the ear dimensions. (5) As the dimensions of the ear decrease, the angle of impedance moves closer to zero and the slope of the magnitude vs. frequency function becomes flatter at the lowest frequencies. The first four effects can be readily explained by simply scaling both the radiation load on the ear and the distributed mass and compliance within the tube (Beranek 1954, Chapter 5; Kinsler and Frey 1962, Chapter 8). As the dimensions of the ear vary, the effective mass of the air driven by the horn at low frequencies varies as one over the tube radius, the open tube resonance frequency increases proportionally with tube length, and the resistive characteristic impedance of the tube varies as one over the radius squared.

The observed changes in impedance angle and magnitude slope at low frequencies, however, can only be due to changes in the relative contribution of tube losses to the impedance. This relative change can be understood if the angles of the different model predictions at 0.2 kHz are compared. At this frequency the impedance angle in the cat-sized model is about 0.24 periods, which indicates that the loss-related resistive component is only 1/20 of the mass-induced reactive impedance. At that same frequency in the down-scaled ear, the 0.125 period impedance angle indicates that the mass reactance and resistive components of the impedance are of equal magnitude. This increase in the relative magnitude of the loss term with decreased dimensions results because the resistance is proportional to the surface area of the tube walls (Egolf 1980; Keefe 1984; Lin 1990), while the reactive components of sound flow in tubes are related to the volume of the tube. The smaller the ear tube, the larger the surface area to volume ratio and the more prominent the loss component.

One elementary method for assessing the quality of power collection by the ear is a simple comparison of the external ear radiation impedance, Z_E, and the middle ear input impedance, Z_T. Circuit theory tells us that a source transfers maximum power to a load when the output impedance of the source and the input impedance of the load are "matched," i.e., of equal magnitude and opposite angle (Desoer and Kuh 1969). To assess impedance matching at the tympanic membrane, consider the external ear radiation impedance, Z_E, as the output impedance of a sound source (Siebert 1970; Shaw 1976, 1988; Rosowski, Carney, and Peake 1988) and the middle ear input impedance, Z_T, as the impedance of the load. Figure 6.11 compares measurements of Z_E and Z_T in humans. Matching is poorest at frequencies below 1 kHz where the two impedances are of greatly different magnitudes and angles. The two impedances come closer to matching above 2 kHz where the magnitudes of the impedances are roughly equal and the angles are either small or of opposite sign. Qualitative comparisons of these impedances in cats (Rosowski, Carney, and Peake 1988) and gerbils (Ravicz 1990) show a similar frequency dependence of matching.

Siebert (1970) and Rosowski (1986, 1988) described a quantitative estimate of matching. This external ear power utilization ratio (PUR) is defined as

$$PUR = \frac{4 \, Re\{Z_E\} \, Re\{Z_T\}}{|Z_E + Z_T|^2}. \tag{2}$$

When Z_E and Z_T match, the PUR = 1, and PUR $\ll$ 1 when matching is poor. Estimates of the PUR for four mammalian ears are displayed in Figure 6.12. Note that in these four species the impedances are only close to matching at selected frequencies above 1.5 kHz. At lower frequencies the large mismatch between the mass-dominated radiation impedance and the stiffness-dominated middle ear input impedance produces PURs of much less than 0.01. The observed lack of perfect matching is contrary to the ideal transformer theory and demonstrates the large effect of the impedance of the external and middle ear structures on power flow through the ear (Peake and Rosowski 1991).

The power utilization ratio is a direct measure of how well the ear collects and delivers power to the tympanic membrane. When the PUR = 1, 100% of the maximum power available enters the middle ear; when the PUR = 0.1,

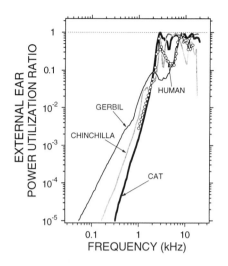

FIGURE 6.12. The external ear power utilization ratio (PUR) estimated for four mammals using the data of Figures 9 and the middle ear impedance data of figure 16. A ratio of 1 (the dashed horizontal line) indicates perfect coupling of power from the external to the middle ear. The estimates at lower frequencies ($f < 2$ kHz in the cat and chinchilla; $f < 6$ kHz in the gerbil) are dependent on horn models of the real part of the radiation impedance. The use of such models is necessitated by errors in the real-part estimate that result from small errors in the measured impedance angles at frequencies where the impedance is dominated by reactive elements (Rosowski, Carney, and Peake 1988).

only 10% of the available power enters the middle ear (Rosowski et al. 1986). The middle and external ears of all four species collect similar percentages (10% to 100%) of the available power at higher frequencies. At frequencies below 1 kHz, the PUR is small for all species, but it is largest in the gerbil.

The acceptance of power at the tympanic membrane can also be described in terms of the "effective area" of an acoustic receiver in the diffuse field (Rosowski 1991; also called the absorption cross-section, Rosowski, Carney, and Peake 1988; Shaw 1988). This area is defined as the ratio of the power that enters the middle ear to the power density (power/area) in the diffuse field (Rosowski, Carney & Peake 1988; Shaw 1988), i.e.

$$EA_{TM}^{DF} = \frac{\text{Power into Middle Ear}}{\text{Power/Unit Area in the Diffuse Field}}$$

$$= \eta_E \frac{\lambda^2}{4\pi} PUR \tag{3}$$

where λ is the wavelength of the sound. The EA_{TM}^{DF} has three terms. The first, η_E, is the efficiency of the external ear which accounts for any power lost to the walls of the auditory tube. In humans and cats, the ear canal losses are

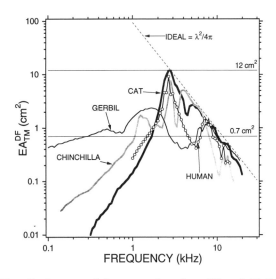

FIGURE 6.13. The effective area of the external ear in a diffuse field, EA_{TM}^{DF}, calculated from Equation 3 using the PUR calculations of Figure 12, with $\eta = 1$. The estimates at lower frequencies (f < 2 kHz in the cat and chinchilla; f < 6 kHz in the gerbil) are dependent on horn model estimates of the real part of the radiation impedance. The dashed diagonal line is the effective area of an ideal spherical receiver. The thin horizontal lines demarcate the area of the pinnae opening in cats (12 cm²) and gerbils (0.7 cm²).

small and η_E approximates 1, at least at frequencies above 1.0 kHz (Shaw 1976; Kuhn 1979; Rosowski, Carney, and Peake 1988). The middle term, λ^2 4π, is an estimate of the power available to the external ear based on power absorption by a sphere placed in a diffuse field. The final term is the PUR, the index of power absorbed over power available at the tympanic membrane. Figure 6.13 illustrates the EA_{TM}^{DF} for four species. The ordinate scale can be interpreted as the area of a perfect receiver required to capture the power that enters the middle ear (Shaw 1988; Rosowski 1991). The thin dashed line defines the ideal performance that occurs when there is no power lost in the external ear ($\eta_E = 1$) and all available power enters the middle ear (PUR = 1). Since we assume $\eta_E = 1$, the proximity of the estimated areas to the ideal is directly related to the proximity of the PUR to 1 (Fig. 6.12). The magnitude of the best EA_{TM}^{DF} in each species is roughly related to the area of the pinna opening (Table 1), being largest in the cat and smallest in the gerbil.

The calculations in Figure 6.13 suggest that the cat external and middle ears collect the most sound power at frequencies between 2 and 4 kHz but are inferior to the other animals at lower frequencies. Below 1 kHz, the gerbil external and middle ears appear to be the best sound collector by far, followed by the chinchilla and then humans. The suggested superiority of the gerbil is consistent with ideas that the ears of desert animals with hyper-trophied tympanic membranes and middle ear cavities are specialized for the reception of low-frequency sound (Legouix and Wisner 1955, Webster 1965; Lay 1972; Webster and Webster 1975; Plassmann and Webster 1992). How-ever, there are several factors that can have a large influence on the compari-son of power collection in these four species:

(1) As described in the caption of Figure 6.12, the calculations of the PUR and EA_{TM}^{DF} at frequencies below 2 kHz are dependent on a model approxi-mation of the real part of the radiation impedance rather than on mea-surements. Errors in the real part estimate (that could be as large as an order of magnitude) produce proportional errors in the PUR and EA_{TM}^{DF} (Eqs. 2 and 3).

(2) Our assumption that $\eta_E \approx 1$ has been only roughly verified in humans (Shaw 1976; Kuhn 1979; Shaw and Stinson 1983) and cats (Rosowski, Carney, and Peake 1988) for sound frequencies above 1.0 kHz, and the analysis of Figure 6.12 points out that $\eta_E \neq 1$ at very low frequencies or at higher frequencies in ears with smaller dimensions. The increase in ear tube losses with decreased size could have a significant effect on the pow-er collected by the gerbil ear at the lowest frequencies. Indeed, this effect could limit low-frequency power collection in all animals with external ears of small cross-section.

(3) The radiation impedance measured in the gerbil was obtained from an excised external ear. Similar measurements on cat ears appear to be inde-pendent of the body and head. The closer proximity of the gerbil head to the pinna opening and the smaller size of the gerbil ear may alter the radiation impedance by as much as a factor of two. Any head and body effects on Z_E will lead to errors in the PUR and EA_{TM}^{DF} computations.

(4) Not all the power collected by the external ear reaches the inner ear, and species-dependent differences in the efficiency of power flow through the middle ear can have large effects on the power available to the inner ear (Shaw and Stinson 1983; Rosowski et al. 1986; Rosowski 1991).

(5) The inner ears of different mammalian species show varying sensitivities to sound power (Rosowski et al. 1986; Rosowski 1991). This variation in sensitivity makes it impossible to predict absolute auditory threshold from simple estimates of power absorption.

3.2.4 The Influences of the Middle Ear

Since all of the estimates of power collection by the external ear depend on both the radiation impedance, Z_E, and the middle ear input impedance, Z_T, it should be apparent that the middle ear load can have a large effect on external ear function. Furthermore, the whole concept of impedance match-ing depends on the relative magnitudes and phase angles of the source and load impedances (Fig. 6.11). The external ear pressure gain is also theoreti-cally dependent on the middle ear load (Siebert 1970, 1973; Shaw 1976, 1988;

Kuhn 1979; Rosowski, Carney, and Peake 1988), and von Bismark (1967) and Sinyor and Laszlo (1973) have demonstrated that changes in the middle ear input impedance produced by varying the pressure in the middle ear cavities of chinchillas and guinea pigs can have large effects on $\mathbf{P_T}/\mathbf{P_{PW}}$.

3.3 Correlations Between Structure and Function

Shaw (1974a, b) addressed the role of several gross structural components in external ear function by estimating the effect of various structural changes in the human ear on external ear pressure gain (Fig. 6.14). These conclusions were later supported by measurements made on mannequins (Kuhn 1977, 1979). The human external ear gain results from the sum of contributions of many structures, including the torso, head, pinna flange, concha, and ear canal. The frequency range over which each structure contributes to the gain depends on the relative size of the structure and the sound wavelength, λ. Structures with dimensions of $<0.1 \lambda$ have little effect. At sound frequencies below 0.6 kHz, the wavelength of sound is greater than 56 cm and the only human structures with dimensions of at least 0.1λ are the torso, neck, and head. At frequencies between 0.8 and 2 kHz, the wavelength of sound is between 44 and 17 cm, and the 3-cm long ear canal becomes important to external ear gain. In this same frequency range, the contribution of the torso and neck decreases as they become separated from the ear by larger and larger fractions of wavelengths. Above 2 kHz, the smaller structures within the ear tube begin to play a role in ear gain. Similar relationships between size and wavelength bound the role of the external ear structures in all mammals, and it can be expected that the largest external ear gains in small animals will be at high frequencies.

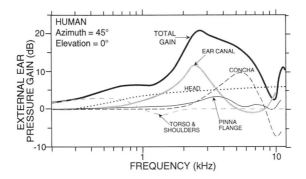

FIGURE 6.14. The contribution of various structures to the external ear gain in humans (after Shaw 1974a). Estimates of the gain are made for a stimulus direction of 45° azimuth and 0° elevation. The total gain is produced by multiplication (dB addition) of all of the other gains.

3.4 Summary of External Ear Measurements

In the last twenty years, measurements of external ear gain and radiation impedance in new species and more sophisticated measurements on standard laboratory mammals have led to better estimates of the role of the external ear in audition, but much work still needs to be done. Direct measurements of the interaction of the head, body, and finer structures of the external ear on both directionality and power collection are needed in more species. Also needed are more measurements of the effects of variations in external ear structure on function. The theoretical approximations of the role of external ear structures can also be improved to yield tractable explanations that can advance the intuition and understanding of the complex auditory role of the external ear.

One important structural quality that effects both directionality and power collection is size. Larger external ear areas are more directional at lower frequencies and make more power available to the middle ear at all frequencies, while long and narrow external ear dimensions can decrease the low-frequency power available to the middle ear. Such simple size rules predict that fossorial mammals that do not localize sounds and are dependent on low-frequency sounds for communication (e.g., the pocket gopher; Heffner and Heffner 1990) would have wide, short ear canals, while those small mammals with narrow ear tubes (e.g., mice and rats) would have poor low-frequency hearing (Fay 1988).

4. Measurements of Middle Ear Structure and Function

Some general ideas of how the structure of the mammalian middle ear affects its function have already been discussed. In this section, I will (1) expand on this topic by describing structural variations, (2) review existing measures of middle ear function in mammals, and (3) use the cascade and power flow descriptions of the middle ear (Figs. 6.3 and 4) to compare middle ear function among mammals.

4.1 Variations In Middle Ear Size and Structure

4.1.1 The Areas of the Tympanic Membrane and Stapes Footplate

The areas of the tympanic membrane, A_{TM}, and stapes footplate, A_{FP}, vary with body size (Khanna and Tonndorf 1978; Hunt and Korth 1980; Rosowski and Graybeal 1991). Estimates based on a sample of 56 terrestrial mammals suggest that the tympanic membrane area grows as the fourth root of body weight (Fig. 6.15A), and a similar relationship is observed between the footplate area and body weight (Hunt and Korth 1980; Rosowski and Graybeal 1991; Rosowski 1992). The similar scaling of the tympanic membrane and footplate areas is also observed in direct comparisons of these two di-

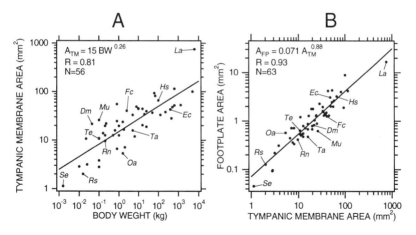

FIGURE 6.15. Empirical relationships of the area of the tympanic membrane with body weight and with the area of the stapes footplate. (A) The relationship between the area of the *pars tensa* of the tympanic membrane and the body weight of 56 terrestrial mammals ranging in size from the dwarf shrew, *Se*, to the African elephant, *La*, is well fit by a power function, A_{TM} (mm^2) = 15 (mm^2/kg) × BW(kg)$^{0.26}$, where the probability that the exponent is 0 is much less than 1%. (B) The relationship between footplate and pars tensa areas observed in 63 terrestrial mammals is also well fit by a power function, A_{FP} (mm^2) = 0.071 A_{TM} (mm^2)$^{0.88}$. The exponent relating the areas is highly significant, but not significantly different from 1. The data are taken from Wever and Lawrence (1954), Kirikae (1960), Lay (1972), Fleischer (1973), Webster and Webster (1975), and Hunt and Korth (1980). A few points in both plots are labeled: *Dm*, kangaroo rat *Dipodomys merriami*; *Ec*, horse *Equus caballus*; *Fc*, cat *Felis catus*; *Hs*, *Homo sapiens*; *La*, elephant *Loxodonta africans*; *Mu*, Mongolian gerbil *Meriones unguiculatus*, *Oa*, platypus *Ornithorhynchus anatinus*; *Rn*, rat *Rattus norvegicus*; *Rs*, bat *Rhinolophus simulator*; *Se*, dwarf shrew *Suncus etruscus*; *Ta*, echidna *Tachyglossus aculeatus*; and *Te*, mole *Talpa europa*. R is the Pearson-product-moment correlation coefficient.

mensions in 63 terrestrial mammals (Fig. 6.15B). A power function fit to the two areas is highly significant, with an exponent near 1, and 78% of the variation in log (A_{TM}) and log (A_{FP}) is explained by a simple linear relationship, A_{TM} = 20 A_{FP} (Rosowski 1992 came to a similar conclusion with a slightly different data base).

The middle ear areas also correlate with differences in the frequency range of hearing sensitivity. Animals with large tympanic membrane and footplate areas usually are better at hearing low-frequency sounds; animals with small middle ear areas usually are better at hearing high frequency sounds (Fleischer 1973, 1978; Rosowski and Graybeal 1991; Rosowski 1992). No correlation is apparent between middle ear areas and hearing thresholds. There is also no evidence of a correlation between best threshold and the ratio of the middle ear areas, A_{TM}/A_{FP}, as would be predicted by the ideal transformer hypothesis (Rosowski and Graybeal 1991).

TABLE 6.2. The tympanic membrane (TM) and stapes footplate areas and the volume of the middle ear cavity in the 13 mammals in which there are middle ear physiological and anatomical data.

Species	Author	Body weight (kg)	Area of pars tensa of TM (mm²)	Area of pars flaccida of TM (mm²)	Area of stapes footplate (mm²)	Effective area of TM $|A_{ET}|$*	Volume of middle ear cavity (cm³)
Bats:							
E. pumilis	Manley, Irvine, and Johnstone 1972		2.1				
Rhinolophus simulator	Fleischer 1973	0.02	2.0		0.23		
Cat							
Felis catus	Fleischer 1973	2.5	36				
	Kohllöffel 1984			0.7		45	
	Guinan and Peake 1967				1.25		
	Lynch 1981						0.89
Chinchilla							
Chinchilla laniger	Fleischer 1973	0.8	71		1.9		
	Vrettakos, Dear, and Saunders 1988		56	0.11	2.0	70	1.5
	Drescher and Eldredge 1974						2.2
	Teas and Nielsen 1975						2.8
Echidna							
Tachyglossus aculeatus	Fleischer 1973	5	15.9		0.67		
Guinea pig							
Cavia procellus	Mundie 1963	0.4	25				
	Fleischer 1973					37	0.25
	Kohllöffel 1984			0	0.88		
	Teas and Nielsen 1975						0.2
	Wilson and Johnstone 1975						0.31
Hamster							
Mesocricetus auratus	Stephens 1972	0.4	7			10	
	Kohllöffel 1984			0			
	Zwillenberg, Konkle, Saunders 1981						0.041
Human	Wever and Lawrence 1954	75	60	1.8	3.2	50	
	Molvær et al. 1978						6.5

Species	Reference					
Kangaroo rat *Dipodomys merriami*	Webster and Webster 1975	0.05	21		0.83	0.47
Mongolian gerbil *Meriones unguiculatus*	Lay 1972	0.1	26	1.6	0.62	0.22
	Ravicz 1990		14			0.20
Mouse *Mus musculus*	Saunders and Summers 1982	0.05	2.8	1.2	0.09	0.05
	Personal observation					
Rabbit *Oryctolagus cuniculus*	Kirikae 1960	1.8	23		0.76	
	Kohllöffel 1984		17		1.3	
	Møller 1965			3		0.13
Rat *Rattus norvegicus*	Fleischer 1973	0.2	9.6		0.43	
	Kohllöffel 1984					
	Ravicz (unpublished)			2.6		0.08

*Approximate low-frequency value of the ratio of the magnitude of TM volume velocity and umbo velocity ($|\mathbf{U}_T/\mathbf{V}_U|$) from Figure 18 in mm^2.

In many mammals, the tympanic membrane consists of two different components (Kohllöffel 1984; Vrettakos, Dear, and Saunders 1988). The membrane component bound to the malleus handle is usually referred to as the *pars tensa* due to its stiff appearance. The second, usually much smaller, "accessory" component is either superior or posterior to the lateral process of the malleus and is called Shrapnell's membrane, or *pars flaccida*, due to its slack appearance in many species (Shrapnell 1832; Kohllöffel 1984; Unge, Bagger-Sjöbäck, and Borg 1991). [In a few species, notably the goat *Capra aegagrus hircus* and pig *Sus scrofa*, the *pars flaccida* is nearly as large or larger than the *pars tensa* (Kohllöffel 1984).] The two membrane components are different in structure; the *pars tensa* has a well-developed fibrous layer that is absent in the *pars flaccida* (Lim 1968; Funnell and Laszlo 1982). The areas of the *pars tensa* and *par flaccida* in a few animals are compared in Table 6.2. The *pars tensa* is the membrane component usually assumed to be important in middle ear sound transmission (Fleischer 1973, 1978; Kohllöffel 1984; Rosowski 1992). The possible function of the *pars flaccida* has been discussed by Kohllöffel (1984), and Vrettakos, Dear, and Saunders (1988) and is reviewed in Section 4.3.8.

4.1.2 Variations in Middle Ear Cavity Volume

Another mammalian middle ear structure that varies widely in size and configuration is the middle ear air space (Pocock 1921, 1928; van der Klaauw 1931; Werner 1960; Hunt 1974; Novacek 1977). The volumes of middle ear air in a few mammals are included in Table 6.2, and even in this small population, the volume varies over several orders of magnitude, from 0.06 cm^3 in the mouse to 6.5 cm^3 in humans. (Intraspecies variation in middle ear air volume can be large, e.g., in 55 human middle ears, the air volume varied between 2 and 22 cm^3, with a mean of 6.5 cm^3 [Molvær, Vallersnes, and Kringlebotn 1978].) The shape of the air spaces also varies greatly, from a simple egg shape with one or two compartments, as in cats and many carnivores (Hunt 1974), through the multiple compartments of the ears of chinchillas, gerbils, and some other rodents (Lay 1972; Vrettakos, Dear, and Saunders 1988) to the spongelike structures of the mastoid air cells that make up much of the volume of the cavities in humans and other large primates (Wever and Lawrence 1954). The bones that surround the air spaces come from as many as eight different sources, including the tympanic, petrous, squamosal, ectotympanic, and endotympanic bones, and the bony makeup of the cavity walls is often used to differentiate phylogenetic sequences (Hunt 1974,1987; Novacek 1977).

4.1.3 Variations in the Size and Structure of the Ossicles

The size, shape, and attachments of the ossicles vary greatly among mammals (Doran 1879; Gray 1913; Henson 1961, 1974). In general, the linear dimensions of the ossicles vary with body size—the larger the mammal, the larger the ossicles. However, this scaling is not universal and some small mammals have huge ossicles (e.g., the Cape golden mole *Chrysochloris trevelyani*; Henson 1974). One of the most variable ossicular structures is the anterior process of the malleus (sometimes called the *process gracilis* or gonial). In adult humans, this process is simply a small nub of bone to which the anterior ligament of the malleus attaches (Bast and Anson 1949; Kobrak 1959); in children, the process appears longer and may even form a short synarthrosis (a joint in which the bony elements are connected by a continuous band of connective tissue) with the tympanic bone (Doran 1879). Carnivores possess a slightly elongated anterior process that attaches to the tympanic bone (Fleischer 1978), while in many small mammals (e.g., the dwarf shrew; Fig. 6.1) the anterior process is greatly expanded and firmly attached to the tympanic bone via a long synarthrosis (Fleischer 1973, 1978; Henson 1974). Within Rodentia are species with practically nonexistent anterior processes, e.g., guinea pigs and chinchillas, as well as species in which the anterior process is both large and firmly attached to the tympanic bone, e.g., the house mouse and lab rat (Cockerell, Miller, and Printz 1914). The most prominent anterior processes seem to occur in bats and shrews (Henson 1961, 1974; Fleischer 1973). Fleischer (1973, 1978) has suggested that the anterior process is a distinguishing functional feature in that mammals with prominent anterior processes tend to be insensitive to sounds at frequencies less than 1 kHz and more sensitive to sounds at higher frequencies. An examination of the behavioral and anatomical evidence from a small number of mammalian species supports this hypothesis (Rosowski 1992).

Associated with variations in the prominence of the anterior process are variations in the orientation of the manubrium of the malleus relative to the long arm of the incus and the horizontal plane (Fleischer 1973, 1978). In mammalian species where the anterior process is nearly absent (e.g., primates, chinchillas, and guinea pigs), the manubrium and long arm of the incus are usually nearly parallel and both are perpendicular to the horizontal plane. In mammalian species with a prominent anterior process (e.g., the dwarf shrew; Fig. 6.1), the manubrium is nearly parallel with the horizontal plane and perpendicular to the vertical long arm of the incus. In ears with an anterior process of intermediate prominence (e.g., many carnivores), the angle of the manubrium is usually intermediate between that of the horizontal plane and the vertical long incus arm (Werner 1960; Fleischer 1973, 1978).

4.2 Measurements of Function

4.2.1 Linearity

The question of linearity is important to any physical system. In a linear system, the output is proportional to the input and the normalized system "transfer function" (the ratio of an output to an input) is independent of level. Measurements of middle ear impedance, ossicular motion, and cochlear sound pressure made in several mammalian species with physiologic stimulus sound levels (after inactivation of the middle ear muscles) strongly suggest that the mammalian middle ear is linear for sound pressures less than about 130 dB SPL (Guinan and Peake 1967; Gundersen 1971; Wilson and Johnstone 1975; Dancer and Franke 1980; Nedzelnitsky 1980; Relkin and Saunders 1980; Buunen and Vlaming 1981; Saunders and Summers 1982). Demonstrations of nonlinearities within the cochlear load (Rhode 1971; Sellick, Patuzzi, and Johnstone 1982; Robles, Ruggero, and Rich 1986) and the presence of cochlear echoes and distortion products in the ear canal (Kemp 1978; Kim 1980) do not disprove middle ear linearity, since the magnitudes of these nonlinear phenomena are usually only a small fraction of the middle ear's linear response (Rosowski, Peake, and Lynch 1984).

4.2.2 Middle Ear Input Impedance

The tympanic membrane is the entrance to the middle ear. The two sound inputs at the entrance, the volume velocity, U_T (the velocity of the tympanic membrane integrated over its area), and the sound pressure, P_T, acting on the lateral surface of the tympanic membrane, are related by the middle ear input impedance, $Z_T = P_T/U_T$. The available measurements of this impedance made in eight mammalian species are compared in Figure 6.16. Although there are interspecies differences of several orders of magnitude, there are also some similarities. The most consistent feature of these data is that they all describe a springlike (stiffness dominated) impedance at low frequencies. This stiffness dominance is apparent in the inverse relationship between impedance magnitude and frequency and the impedance angles of about -0.25 periods (Fletcher and Thwaites 1979). At higher frequencies, the impedance is resistancelike, with magnitude independent of frequency and the angle ≈ 0. The transition between springlike and resistancelike is sometimes designated as the "middle ear resonance," though this is really a misnomer. The term resonance suggests an impedance minimum produced by the cancellation of negative (springlike) and positive (masslike) reactances. The patterns observed in Figure 6.16 are more like those of high-pass filters produced by the parallel combination of a spring and a mechanical resistance. The frequency of the spring resistance transition is lowest in humans and largest in rats and rabbits. (The hamster data do not cover a wide enough frequency range to define a resistancelike range.) Despite the common view that the mass of the ossicles and tympanic membrane should dominate the middle ear response at high frequencies (e.g., Møller 1965; Shaw and Stinson 1983), the magnitude and angle data indicate there is no obvious mass-controlled frequency range in the impedance, except the very narrow range near 4 kHz in the cat where the middle ear cavities resonate (Møller 1965; Guinan and Peake 1967; Lynch, Peake, and Rosowski 1993). (An acoustic mass would produce an impedance magnitude which is proportional with frequency and an angle of 0.25 periods.)

The middle ear impedance uses volume velocity, U_T, as a measurement of average tympanic membrane motion and holographic measurements of this motion (e.g., Khanna and Tonndorf 1972) clearly demonstrate that the tym-

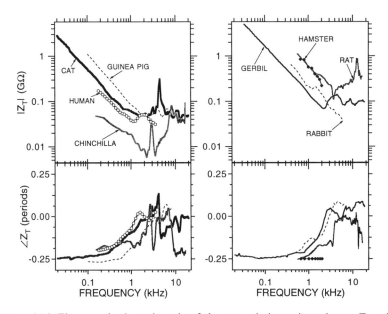

FIGURE 6.16. The magnitude and angle of the acoustic input impedance, Z_T, of the ears of the eight mammalian species in which measurements are available. With the exception of the cat, all measurements were made with the middle ear cavities intact but vented. The data come from different sources. Cat *Felis catus*: mean of the open cavity middle ear input impedance measured in five ears, corrected for the effects of the middle ear cavity in one ear (Rosowski et al. 1986). Chinchilla *Chinchilla laniger*: mean of measurements made in four ears by Dear (1987), corrected after Rosowski (1991). Hamster *Mesocricetus auratus*: mean of measurements made in eight ears from 150-day-old hamsters by Zwillenberg, Konkle, and Saunders (1981). Human: mean of measurements made in four awake subjects by Rabinowitz (1981). Gerbil *Meriones unguiculatus*: mean of measurements made in four ears by Ravicz, Rosowski, and Voigt (1992). Guinea pig *Cavia procellus*: mean of measurements made in 10 ears by Mundie (1963). Rabbit *Oryctolagus cuninculus*: one ear from Møller (1965). Laboratory rat *Rattus norvegicus*: one ear by Ravicz, Rosowski, and Hall (unpublished). The interspecific variation is much larger than the variation within species. One acoustic GΩ equals 10^9 Pa-s-m^{-3}.

panic membrane does not move as a plate. With low stimulus frequencies, there are membrane regions that move with much greater magnitude than others, while at higher frequencies different membrane regions move with different stimulus phases. Again, these details are averaged by the acoustic impedance measurements.

4.2.3 Motion of the Malleus

During sound stimulation, the regions of the tympanic membrane that are coupled to the malleus tend to move less than the surrounding regions

(Khanna and Tonndorf 1972). The decrease in motion results because the malleus, together with the attached ossicles and cochlea, loads the parts of the membrane that are "coupled" to the manubrium but have less effect on more "distant" parts of the membrane (Funnell and Laszlo 1977; Shaw and Stinson 1983; Rabbitt and Holmes 1986; Funnell, Decraemer, and Khanna 1987). The attachment of the malleus to the tympanic membrane varies among species (Funnell and Laszlo 1982). In cats, the malleus appears firmly connected to the membrane along much of its length. In humans, the malleus and tympanic membrane are firmly connected only at the umbo (the tip of the manubrium) and the lateral process of the manubrium (near the tympanic ring) where the malleus is embedded tightly in the fibrous layers of the membrane (Schuknecht 1974; Graham, Reams, and Perkins 1978).

The umbo is a convenient place to measure mallear motion, and the relationship between sound pressure at the tympanic membrane and umbo velocity, V_U/P_T, is illustrated for 10 available species in Figure 6.17. Interspecies differences of several orders of magnitude are apparent; at 1 kHz, the magnitude of V_U/P_T in the chinchilla is 1 mm-Pa^{-1}-s^{-1} while the $|V_U/P_T|$ in echidna is less than 0.01 mm-Pa^{-1}-s^{-1}. In general, the decreased low-frequency umbo velocities apparent in the five species on the right are indicative of increased stiffness in those ears.

Since both Z_T and V_U/P_T quantify tympanic membrane motion, one might expect them to be related. As in the impedance measurements, the low-frequency umbo velocity measurements are all consistent with a springlike mechanism; $|V_U/P_T|$ increases proportionately with frequency and $\angle(V_U/P_T)$ is near 0.25 periods. (The input impedance, Z_T, and the tympanic membrane transfer function, V_U/P_T, are *inversely* related, and springlike behavior in the two have opposite angles and frequency dependence.) Comparisons of Figures 6.16 and 6.17 point out the inverse relationship between V_U/P_T and acoustic impedance; animals with small $|Z_T|$s generally have large $|V_U/P_T|$s. Both the impedance and the umbo velocity magnitude flatten out in the mid- and high-frequency regions. Interestingly, the magnitudes of the mid- to high-frequency velocity plateau of all but the echidna fall within a decade range of 2 to 0.2 mm-Pa^{-1}-s^{-1}, and, because of this similarity, the transition frequency between springlike and resistancelike responses varies inversely with the magnitude of the low-frequency response (ears with larger low-frequency umbo velocities have a lower transition frequency).

There are suggestions that Z_T and V_U/P_T are no longer related at the highest measured frequencies. The angle of Z_T is approximately 0 between 3 and 10 kHz in the seven species for which there are measurements (Fig. 6.16) while, in the two ears where it is measured, $\angle(V_U/P_T)$ becomes increasingly more negative as the frequency increases from 2 to 10 kHz (Fig. 6.17). The magnitude of V_U/P_T also seems to decrease at the highest frequencies where $|Z_T|$ is approximately constant. However, comparisons of measurements made in the same species but different laboratories don't always show similar patterns of high-frequency motion. For example, measurements of the mag-

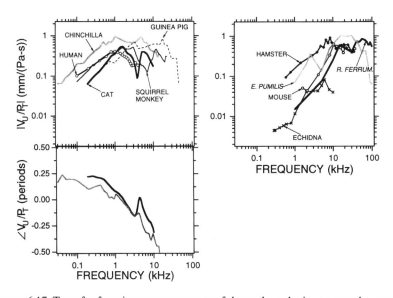

FIGURE 6.17. Transfer-function measurements of the umbo velocity to sound pressure at the TM ratio $\mathbf{V}_U/\mathbf{P}_T$ in 10 mammalian species in which measurements are available, with the middle ear cavity intact but vented (except where specified). Those animals with large velocity, low-frequency velocity responses are plotted in the left panels. Angle measurements are only available for the cat and chinchilla. Cat: mean of measurements in three ears made by Buunen and Vlaming (1981) and in another three ears by Decraemer, Khanna, and Funnel (1989). The sharp dip in umbo motion near 4 kHz again is a sign of the middle ear cavity resonance described by Møller (1965) and Guinan and Peake (1967). Chinchilla: mean of five ears (Ruggero et al. 1990). A bat *Eptesicus pumilis*: mean from five ears measured by Manley, Irvine, and Johnstone (1972). Echidna *Tachyglossus aculeatus*: mean of measurements of stapes motion made in four ears with the middle ear cavity opened and corrected by a measurement of the ratio of umbo and stapes motion made in one ear (Aitkin and Johnstone 1972). Hamster: mean from six ears of animals of 75 days of age (Relkin and Saunders 1980). Human: mean of postmortem measurements made in 14 human temporal bones (Gyo, Aritomo, and Goode 1987). Monkey *Saimiri sciureus*: mean from 11 ears (Rhode 1978). Mouse *Mus musculus*: mean of measurements in 11 ears (Saunders and Summers 1982). The horseshoe bat *Rhinolophus ferrumequinum*; mean of measurements from seven ears (Wilson and Bruns 1983). The interspecific variation is generally much larger than the variation within species.

nitude of umbo velocity in cats made by Johnstone and Taylor (1971) look flat at frequencies out to 50 kHz, while the measurements made by Buunen and Vlaming (1981) and Decraemer, Khanna and Funnell (1989) all roll off precipitously above 10 kHz.

As the cascade model of Figure 6.3 suggests, the relationship between $\mathbf{Z}_T$ and $\mathbf{V}_U/\mathbf{P}_T$ can be described by an "effective area" equal to one over the product of $\mathbf{V}_U/\mathbf{P}_T$ and $\mathbf{Z}_T$ (Fig. 6.3), i.e.,

$$A_{ET} = \frac{1}{Z_T(V_U/P_T)} = \frac{U_T P_T}{V_U P_T} = \frac{U_T \, (mm^3)}{V_U \, (mm)} \tag{4}$$

where A_{ET} has units of area. We can calculate A_{ET} for the cat, chinchilla, guinea pig, hamster, and human using the data from Figures 6.16 and 6.17. While the ideal transformer model predicts that the tympanic membrane moves as a rigid piston, with A_{ET} as a real number (angle of zero) and with

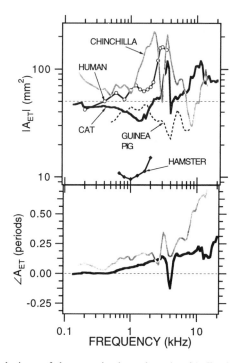

FIGURE 6.18. Calculations of the magnitude and angle of "effective area" of the tympanic membrane, A_{ET}, in the cat, chinchilla, guinea pig, hamster, and human, based on the measured middle ear input impedance and umbo motion and Equation 4. Note that this effective area is a complex, frequency-dependent quantity with both a magnitude and angle (though there are no angle data available for the guinea pig, hamster, and human). The estimate of A_{ET} for the cat was calculated from the Z_T of Lynch (1981) and the V_U/P_T data of Buunen and Vlaming (1981). The fine structure in both the magnitude and angle of A_{ET} of the cat near 4 kHz results from differences in the sharpness of the cavity resonance in the two measurements. The chinchilla A_{ET} was estimated from Dear's (1987) measurements of Z_T and Ruggero et al.'s (1990) measurements of V_U/P_T. The guinea pig A_{ET} was calculated from the data of Mundie (1963) and Manley and Johnstone (1974). The hamster A_{ET} was calculated from the data of Zwillenberg, Konkle, and Saunders (1981) and Relkin and Saunders (1980). The human A_{ET} was calculated from the data of Rabinowitz (1981) and the measurements of Gyo, Aritomo, and Goode (1987). The dotted horizontal lines describe an A_{ET} of zero angle and magnitude of 50 mm².

the magnitude equal to the actual area of the membrane, calculations of A_{ET} based on Equation. 4 (Fig. 6.18) indicate that A_{ET} is both frequency dependent and complex (i.e., V_U and U_T are not in phase and the angle is nonzero). Indeed, A_{ET} is only close to real at the lowest frequencies where all portions of the tympanic membrane appear to move in phase (Khanna and Tonndorf 1972).

Comparisons of the anatomical area of the TM and the magnitude of A_{ET} at low frequencies are made in Table 6.2. Since the holographic results (Khanna and Tonndorf 1972; Tonndorf and Kharma 1972) indicate that much of the tympanic membrane moves with larger amplitudes than the umbo, it is not surprising that $|A_{ET}|$ is actually *larger* than the anatomical area in four of the five ears where this comparison is made. Humans are the one species where the low-frequency data suggest that the "effective area" is substantially less than the anatomical area of the tympanic membrane. The low-frequency calculations from the human data suggest that $|A_{ET}|$ is about 80% of A_{TM}, somewhat larger than the 66% estimated by Wever and Lawrence (1954). The variations in A_{ET} magnitude and angle at higher frequencies reflect frequency-dependent variations in the pattern of tympanic membrane and umbo motions (Khanna and Tonndorf 1972; Tonndorf and Khanna 1972).

There are several complications in the concept of "effective area."

(1) Recent measurements of motion along the arm of the malleus in the cat have suggested that (a) the mallear arm is not rigid (Funnel, Decraemer, and Khanna 1992), and (b) the motion of the malleus is made up of several components including translation and rotation (Decraemer, Khanna, and Funnel 1991). Similar findings have been described in a human temporal bone preparation (Gyo, Aritomo, and Goode 1987; Hüttenbrink 1988; Donahue, Rosowski, and Peake 1991). Any bending or complex motion of the malleus greatly complicates the relationship between tympanic membrane and umbo motions.

(2) Since the tympanic membrane does not move as a rigid piston, the magnitude of the "effective area" is entirely dependent on how the area is defined. Simple circuit theory suggests that an effective area based on the ratio of *force* and pressure, $A_{EM}^{FORCE} = F_U/P_T$, does not necessarily equal the effective area that has been defined from measurements of *velocity*, $A_{ET}^{VELOCITY} = U_T/V_U$ (Shera and Zwieg 1991).

4.2.4 Motion of the Stapes Footplate

The middle ear ossicular system converts the motions of the umbo and malleus into motion of the stapes, and the next stage of the cascade model relates umbo and stapes motions via the middle ear velocity transfer ratio, V_S/V_U. The ideal transformer model assumes simple rotation of the ossicles and predicts that V_S/V_U would be real (the motions of the umbo and stapes in phase), independent of frequency, and equal to the ratio of the lengths of the

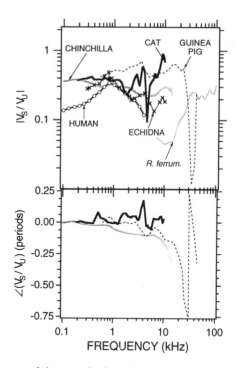

FIGURE 6.19. Estimates of the magnitude and angle of the ossicular velocity transfer ratio, $\mathbf{V_S}/\mathbf{V_U}$. The data used in the calculations are: cat— the umbo velocity measurements of Buunen and Vlaming (1981) and the stapes motion measurements of Guinan and Peake (1967); chinchilla—Ruggero et al. (1990); echidna—Aitkin and Johnstone (1972); guinea pig—Manley and Johnstone (1974); human—the umbo velocity measurements of Gyo, Aritomo, and Goode (1987) and the stapes motion measurements of Kringlebotn and Gundersen (1985); and horseshoe bat *Rhinolophus ferrumequinum* —Wilson and Bruns (1983). The measured ratios are complex and frequency dependent. The ideal transformer model predicts a real (angle = 0) velocity ratio equal to the ratio of the ossicular lever arms, L_I/L_M, which is about 0.5 in cat and 0.8 in humans.

incus and malleus lever arms, L_I/L_M. Measurements of this ratio from five mammalian species (Fig. 6.19) show that $\mathbf{V_S}/\mathbf{V_U}$ varies with frequency in both magnitude and angle at frequencies above 2 kHz. These variations indicate that the ossicles do not simply rotate. Complex patterns of motion, bending of the ossicles, or compression and flexion of the ligaments that connect the ossicles could all contribute to the observed deviations from rotation. At frequencies below 2 kHz, $\mathbf{V_S}/\mathbf{V_U}$ in the cat, chinchilla, guinea pig, and echidna best approximate simple rotation in that $\angle(\mathbf{V_S}/\mathbf{V_U})$ is about 0 and $|\mathbf{V_S}/\mathbf{V_U}|$ approximates L_I/L_M (about 0.5 in the cat), suggesting that low-frequency motion of the umbo is usually transmitted to the stapes without losses due to the compression of the ossicular joints, bending of the ossicles, or large changes in ossicular motion. The systematic variations from constant magnitude and zero angle in a few species at higher frequencies (cat, Guinan and Peake 1967; guinea pig, Manley and Johnstone 1974; chinchilla, Ruggero et al. 1990) indicate that nonideal ossicular motions are more significant above 5 kHz. The low-frequency fall-off observed in the human data (Fig. 6.19) is consistent with the observations of Hüttenbrink (1988) who suggested that compressible ossicular joints, ossicular bending, and changes in the mode of

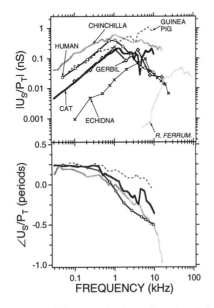

FIGURE 6.20. Measurements of the magnitude and angle of the middle ear transfer function, U_S/P_T (stapes volume velocity per sound pressure at the TM), in the seven mammalian species in which measurements are available. All measurements were of stapes velocity made with the middle ear cavity opened; however, the data have been corrected for opening the middle ear air space. In order to convert the measurements to volume velocity, a pistonlike motion of the stapes was assumed (Guinan and Peake 1967; Gyo, Aritomo, and Goode 1987) and the measured velocities were simply multiplied by the footplate area. Cat—mean of measurements in 25 cat ears, $A_{FP} = 1.25$ mm^2 (Guinan and Peake 1967). Chinchilla—mean of measurements in seven ears, $A_{FP} = 2.0$ mm^2 (Ruggero et al. 1990). Echidna—mean of measurements in four ears, $A_{FP} = 0.67$ mm^2 (Aitkin and Johnstone 1972). Gerbil— calculated from the cochlear potential measurements of Woolf and Ryan (1984, 1988) and corrected for opening the middle ear cavity using the middle ear input impedance data of Ravicz (1990), $A_{FP} = 0.62$ mm^2. Guinea pig—mean of measurements in four ears, $A_{FP} = 0.88$ mm^2 (Nuttall 1974). Human—mean of round-window volume velocity measurements made in over 30 human temporal bones with the middle ear cavities intact (Kringlebotn and Gundersen 1985). Horseshoe bat *R. ferrumequinum*—mean of measurements in five ears, $A_{FP} = 0.12$ mm^2 (Wilson and Bruns 1983). One nS equals 10^{-9} siemens equals 10^9 m^3·Pa^{-1}·s^{-1}.

ossicular motion may play a role in protecting the human ear from large static pressures and intense low-frequency sounds.

The pistonlike velocity of the stapes (Guinan and Peake 1967; Gyo, Aritomo, and Goode 1987) is converted into volume velocity of the footplate, where $U_S = V_S A_{ES}$ and the effective area of the stapes footplate is usually approximated by the anatomical area of the stapes footplate, $A_{ES} \approx A_{FP}$. The ratio of stapes volume velocity to sound pressure at the tympanic membrane, U_S/P_T, is one measure of the overall middle ear function. The ratio U_S/P_T has units of an acoustic admittance (the inverse of impedance, where 1 siemen = 1 m^3-Pa^{-1}-s^{-1}) and is sometimes referred to as the middle ear transfer admittance. Measurements of the transfer admittance in seven mammalian species are compared in Figure 6.20. These ratios are generally of similar shape to the umbo velocity measurements; in all ears, the stapes

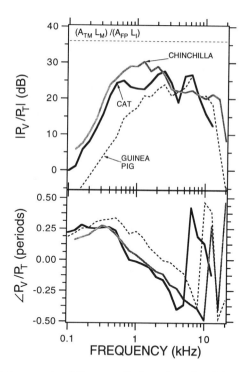

FIGURE 6.21. Measurements of the magnitude and angle of the middle ear pressure gain function, P_V/P_T, in three mammalian species (after Décory 1989). Measurements made with the middle ear open in eight cat, nine chinchilla, and 11 guinea pig ears have been corrected for the effects of opening the middle ear using the data of Guinan and Peake (1967), Nuttall (1974), and Ruggero et al. (1990). The dashed horizontal line illustrates, the middle ear pressure gain predicted from anatomical measurements of the tympanic membrane and footplate areas and malleus and incus lengths of the cat.

velocity appears stiffness controlled at lower frequencies and the magnitude of the velocity flattens out at higher frequencies. The data from human, gerbil, and echidna ears also show a decrease in magnitude at the highest frequencies. Similar decreases in stapes velocity have been observed in the cat and guinea pig at frequencies above 10 kHz (Guinan and Peake 1967; Manley and Johnstone 1974). The rightward displacement of the echidna and horseshoe bat transfer admittance in Figure 6.21 is another indication that these ears are stiff compared to the others in the measurement population.

4.2.5 Sound Pressure in the Vestibule

Another estimate of the middle ear output is the middle ear pressure transfer ratio, $\mathbf{P}_C/\mathbf{P}_T$. Miniature pressure transducers with tiny probe tubes are placed within the perilymph of the vestibule to measure the sound pressure that results from controlled stimulation of the tympanic membrane. The ratio of these pressures has been measured in the cat by Nedzelnitsky (1974, 1980), in the guinea pig by Dancer and Franke (1980), and in the chinchilla, cat, and guinea pig by Décory (1989). Décory's measurements (adjusted to include the effect of the middle ear cavity) are illustrated in Figure 6.21. Between 0.1 and 0.4 kHz, the pressure ratio measured in all three species grows at a rate near 20 dB per decade and has an angle near 0.25 periods. In the mid-frequency range, $|\mathbf{P}_C/\mathbf{P}_T|$ is relatively flat but $\angle(\mathbf{P}_C/\mathbf{P}_T)$ becomes increasingly more negative. The rate of the angle change appears to increase with increasing frequency much like the angle of the stapes volume velocity measurements (Fig. 6.20). At the highest frequencies, the magnitude of $\mathbf{P}_C/\mathbf{P}_T$ rolls off steeply. The magnitude roll-off and the increasing angle difference at high frequencies may reflect the relative motions between the ossicles; however, the frequency of the roll-off is dependent on the precise location of the pressure probe within the vestibule (Dancer and Franke 1980).

The ideal transformer model predicts that $\mathbf{P}_C/\mathbf{P}_T$ is independent of frequency, with a magnitude equal to the product of the area and lever arm ratios of the middle ear and an angle of zero, $\mathbf{P}_C/\mathbf{P}_T \approx (A_T/A_{FP})(L_M/L_I)$ (Eq. 1). The observed frequency dependence of the measured $\mathbf{P}_C/\mathbf{P}_T$, together with its nonzero angle, are more arguments against this simple model. Even in the frequency range where $|\mathbf{P}_C/\mathbf{P}_T|$ is approximately constant (0.3 to 10 kHz), the measured magnitude is at least 6 dB smaller than that predicted by the anatomical ratios and the angle varies between 0.25 and -0.5 periods.

4.2.6 Cochlear Input Impedance: The Load on the Middle Ear

The ratio of the sound pressure in the vestibule and the volume velocity of the stapes is the cochlear input impedance, $\mathbf{Z}_C = \mathbf{P}_C/\mathbf{U}_S$, the load on the middle ear. This load reflects the impedances of the fluid-filled spaces of the inner ear, the round window, and the basilar membrane (Zwislocki 1965, 1975; Lynch, Nedzelnitsky, and Peake 1982; Puria and Allen 1991). Measurements of $\mathbf{Z}_C$ in three species are illustrated in Figure 6.22, along with model

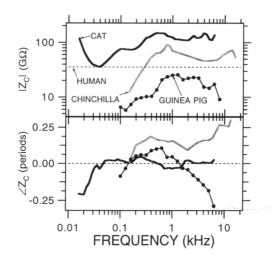

FIGURE 6.22. Estimates of the magnitude and angle of cochlear input impedance, $\mathbf{Z_C}$, in four mammalian ears. Cat—mean of 29 measurements (Lynch, Nedzelnitsky, and Peake 1982). Chinchilla—calculated by Ruggero et al. (1990) from their measurements of $\mathbf{U_S}$ and Décory's (1989) measurements of $\mathbf{P_V}$. Guinea pig—calculated from Nuttall's (1974) measurements of $\mathbf{U_S}$ and Décory's (1989) measurements of $\mathbf{P_V}$. Human—the middle frequency calculation of Zwislocki (1975). The impedance predicted by Zwislocki has been supported by a few measurements in human temporal bones (Aritomo and Goode 1987, Merchant, Ravicz, and Rosowski 1992).

predictions of the impedance of the human cochlea. These measurements suggest that the cochlea is resistancelike (roughly constant magnitude and an angle between ± 0.125) over most of the measurement range. The cat data also show a significant stiffness at lower frequencies (possibly the stiffness of the round window; Lynch, Nedzelnitsky, and Peake 1982). At the high frequencies, the chinchilla angle data suggest a masslike impedance and the angle of the guinea pig impedance looks springlike. Some of the features of the chinchilla and guinea pig $\mathbf{Z_C}$ data may be artifacts induced by combining measurements made in different animals and laboratories.

Manipulations of the cochlea can affect the cochlear impedance; draining the scalae or removing the cochlear partition in the basal regions of the cochlea greatly reduce $\mathbf{Z_C}$ (Lynch, Nedzelnitsky, and Peake 1982). These same manipulations can have a large effect on measurements of middle ear function. For example, Møller (1965), Lynch (1981), Allen (1986), and Tonndorf and Pastici (1986) have demonstrated that removing the cochlear load causes large changes in the measured middle ear input impedance at frequencies near 2 kHz in the cat. These changes suggest that much of the resistance that serves to flatten the cat middle ear impedance and velocity profiles near 2 kHz comes from the cochlea.

4.2.7 Measures of Middle Ear Power Flow: Efficiency and Power Collection

How is middle ear performance ranked? Interspecies comparisons of either middle ear transfer function, U_S/P_T or P_C/P_T, are difficult to interpret without some appreciation of the other, e.g., the P_C/P_T in the cat is larger than that in the guinea pig, and the guinea pig U_S/P_T is larger than that in the cat. One unambiguous method is to estimate the efficiency of the middle ear. The middle ear efficiency (MEE) in ears with known Z_T, U_S/P_T, and Z_C can be estimated (Rosowski 1991), where

$$\text{MEE} = \frac{\text{Power into Inner Ear}}{\text{Power into Middle Ear}}$$

$$= \left|\frac{U_S}{P_T}\right|^2 \frac{|Z_T|^2 \operatorname{Re}\{Z_C\}}{\operatorname{Re}\{Z_T\}}. \tag{5}$$

(The MEE can also be estimated from measurements of Z_T, P_C/P_T, and Z_C). Estimates of the MEE for four species are illustrated in Figure 6.23. The MEE is less than 0.7 in the guinea pig ear at all frequencies and less than 0.3 in the cat, human, and chinchilla. Generally, more than half of the sound power that enters the four middle ears is absorbed there and only the remaining fraction is transmitted to the inner ear. The superior performance of the guniea pig middle ear suggests some special adaptation but requires further study. The data also suggest that a greater power loss occurs at lower and higher frequencies, such that the middle ear efficiency acts as another fre-

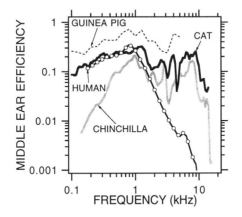

FIGURE 6.23. The middle ear efficiency, MEE. The efficiency of the chinchilla, cat, and human middle ears were calculated, as described by Rosowski (1991). The efficiency of the guinea pig ear was calculated from the U_S/P_T and Z_C data of Figures 20 and 22 and the middle ear input impedance data of Rosowski and Davis (unpublished; these impedance data are similar to those of Mundie (1963) except at the lowest frequencies where the newer data are more springlike and have positive real parts).

quency-dependent filter of sound power that shapes the spectrum of sound signals as they pass through the middle ear to the inner ear (Rosowski 1991).

How does the efficiency of the middle ear affect the overall function of the external and middle ears? A measure of the power transmitted to the inner ear by the combination of the external and middle ears is the effective area from the diffuse field to the oval window, EA_{OW}^{DF}. This area can be calculated from the product of the middle ear efficiency and the effective area from the diffuse field to the TM,

$$EA_{OW}^{DF} = MEE\ EA_{TM}^{DF} \qquad (6)$$

(Shaw and Stinson 1983; Rosowski 1991). Estimates of EA_{OW}^{DF} for three species are illustrated in Figure 6.24. Note the poor power collection performance of these ears when compared to the ideal performance of a spherical collector in a diffuse field. The closest approximation to the ideal performance occurs between 2 and 10 kHz where the cat external and middle ear performance comes within 20% of the ideal. At frequencies below 0.5 kHz, only a tiny fraction (<0.001) of the power available to the ear is collected and passed on to the inner ear. The band-pass shapes of these power filter functions correlate well with the shape of the audiograms in these species (Shaw and Stinson 1983; Rosowski 1991) and are a direct contradiction of the ideal transformer model of middle ear function. These band-pass shapes are directly attributable to the two terms on the right-hand side of Equation. 6.

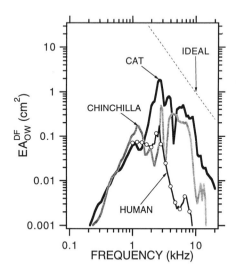

FIGURE 6.24. The effective area from the diffuse field to the oval window, EA_{OW}^{DF}, a measure of the diffuse field sound power that is transmitted to the inner ear. Estimates in three species. The thin dashed line ($\lambda^2/4\pi$) denotes the performance of an ideal spherical receiver (after Rosowski 1991).

(1) The effective area at the tympanic membrane, EA_{TM}^{DF}, describes how most of the power available in sounds at frequencies less than 1 kHz never enters the middle ear but is instead either reflected back into the environment or absorbed within the external ear (Figs. 6.12 and 6.13).

(2) The middle ear efficiency, MEE, describes how much of the power that enters the middle ear is absorbed there (probably by viscous forces within the tympanic membrane and ossicular ligaments or the mucosal lining of the middle ear cavity) and never reaches the inner ear (Fig. 6.24).

4.3 Correlation between Middle Ear Structure and Function

The relationship between middle ear structure and performance is not well understood, but there are a few middle ear structure function features that have been discussed at length in the literature.

4.3.1 The Middle Ear Areas

The ideal transformer model predicts a direct correlation between the area of the tympanic membrane piston and the power available to the middle ear. However, neither the area of the tympanic membrane nor the ratio of the middle ear areas, A_{TM}/A_{FP}, in 16 mammalian species correlates with hearing sensitivity (Rosowski and Graybeal 1991). Rosowski and Graybeal (1991) did find strong correlations between the middle ear areas and the frequency range of best hearing, and Khanna and Tonndorf (1978) have put forward some suggestions on the mechanisms by which the dimensions of the tympanic membrane could limit high-frequency response:

(1) Interactions of receiver dimensions and sound wavelength. To a first-order approximation, the tympanic membrane averages differences in sound pressure acting on its surface. At frequencies where the wavelengths of sound are comparable to or smaller than the membrane dimensions, different parts of the membrane move with different phases (Khanna and Tonndorf 1972) and the average motion is reduced.

(2) Loss of the "coupling" between the tympanic membrane and the manubrium. The holographic data of Khanna and Tonndorf (1972) also pointed out that the membrane motion becomes heterogeneous at high frequencies. At these frequencies, parts of the membrane appear to move independently from the ossicles. Khanna and Tonndorf (1978) suggested that larger membranes exacerbate this effect, causing larger "uncoupled" motions at lower frequencies.

(3) Increases in tympanic membrane area without parallel changes in footplate area could lead to "overmatching" a condition in which the magnitude of a transformer is too large. Overmatching produces power reflection just as undermatching does (Tonndorf and Khanna 1967).

More work needs to be done to test these suggestions. Measurements of the power-collecting function of middle ears of widely varying dimensions and functional descriptions of the frequency dependence of the "effective areas" (Shera and Zwieg 1991) of the tympanic membrane in ears with a wide range of middle ear dimensions are essential.

4.3.2 Stiffness and Middle Ear Size

The stiffness of the middle ear, K_T, plays a large role in defining middle ear function at frequencies below the spring-resistance transition frequency (Figs. 6.16, 6.17 and 6.20). In this low-frequency range,

$$Z_T \approx -j\frac{K_T}{2\pi f} \tag{7}$$

where j is $\sqrt{-1}$ and f is the sound frequency. There is a strong relationship between stiffness and middle ear size, where ears with small tympanic membrane areas and middle ear air volumes tend to be stiffer (Figs. 6.25 and 6.26, Table 6.3). The relationship between total middle ear stiffness and tympanic membrane area is especially significant (Fig. 6.25). The power function fit to these data suggests that the stiffness is inversely proportional to the square of

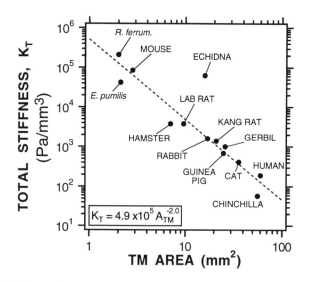

FIGURE 6.25. The relationship between the total acoustic stiffness of the middle ear input impedance, K_T, in Pa-(mm)$^{-3}$ and the area of the tympanic membrane in 13 mammalian species. The stiffnesses are extracted from the low-frequency data of Figures 16 and 17 and are tabulated with the areas in Table 3. The dashed line is a power function fit to the data by least squares regression techniques The exponent of the fitted function is highly significant; the probability that the exponent is zero is less than 0.1%.

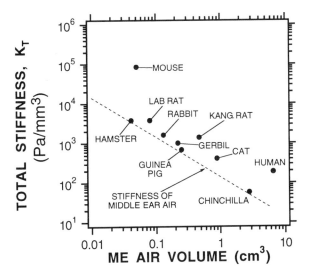

FIGURE 6.26. The relationship between the total acoustic stiffness of the middle ear input impedance, K_T, in Pa-(mm)$^{-3}$ and the volume of the middle ear air space. The stiffnesses are tabulated in Table 3; the middle ear volumes are tabulated in Table 2. The thin dashed line is the theoretical acoustic stiffness of an air space of varying volume, where $K_{CAV} = \rho_0 c^2 \cdot \text{Volume}^{-1}$ (Eq. 9).

the area of the tympanic membrane, and the probability that the exponent is 0 is less than 0.1%. The structural basis for the observed relationship between total stiffness and area is complicated, since the total stiffness results from the contribution of several middle ear stiffnesses (Ravicz, Rosowski, and Voigt 1992), each correlated with middle ear size (Figs. 6.26 and 6.27). It is also noteworthy that this relationship between stiffness and size is peculiar to mammals; reptiles and birds have TM areas roughly equivalent to those of small mammals, but the ears are much less stiff (Rosowski 1992).

4.3.3 The Middle Ear Air Spaces

More than a few authors have suggested a qualitative relationship between large middle ear air volumes, low middle ear stiffness, and increased sensitivity to low-frequency sounds (Legouix and Wisner 1955; Webster 1965, 1982; Lay 1972; Webster and Webster 1975; Zwillenberg, Konkle, and Saunders 1981). The proposed mechanism for this relationship is the "series" impedance model of the cavities (Onchi 1961; Ravicz, Rosowski, and Voigt 1992) that quantifies the relationship between middle ear air volume and motion of the tympanic membrane and ossicles (Zwislocki 1962, 1963; Guinan and Peake 1967; Lynch 1981; Peake, Rosowski, and Lynch 1992). The series model notes that sound-induced motion of the tympanic membrane alternately compresses and rarifies the air within the middle ear air spaces, creating a

TABLE 6.3. The stiffness of the total middle ear and its components in the 13 mammals in which stiffness and tympanic membrane area data are available.

Species	Area of pars tensa (mm²)	Total† acoustic stiffness of the middle ear, K_T (Pa-mm^{-3})	Acoustic†† stiffness of the middle, ear air spaces, K_{CAV} (Pa-mm^{-3})	Acoustic††† stiffness of TM and ossicles, K_{OC} (Pa-mm^{-3})	Stiffness ratio K_{OC}/K_T (%)
Horseshoe bat *Rhinolophus ferrumequinum*	2.0*	210000** Wilson and Bruns 1983	—	—	—
Mouse *Mus musculus*	2.8	85000** Saunders and Summers 1982	3000	82000	96
Echidna *Tachyglossus aculeatus*	15.9	62000** Aitkin and Johnstone 1972	—	—	—
Bat *E. pumilis*	2.1	42000** Manley, Irvine, and Johnstone 1972	—	—	—
Laboratory rat *Rattus norvegicus*	9.6	3800 Ravicz, Rosowski, and Hall 1990	1800	2000	53
Golden hamster *Mesocricetus auratus*	7.0	3800 Zwillenberg et al. 1981	3400	400	11
Rabbit *Oryctolagus cuniculus*	17.1	1600 Møller 1965	1100***	500	31
Kangaroo rat *Dipodomys merriami*	21	1400****	300	1100	79
Gerbil *Meriones unguiculatus*	26	1000 Ravicz, Rosowski, and Voigt 1992	640	360	36
Guinea pig *Cavia procellus*	25	680 Mundie 1963	560	120	18
Cat *Felis catus*	36	410 Lynch, Peake, and Rosowski 1994	160	250	61

Human	60	190	21	170	89
		Rabinowitz 1981			
Chinchilla	56	58	50	8	14
Chinchilla laniger		Dear 1987			

† Acoustic stiffness was defined from low-frequency (less than or equal to 1 kHz) measurements of middle ear input impedance. Estimates of stiffness may vary depending on the frequency over which the estimate was made.

†† $K_{CAV} = \rho c^2/$volume of the cavity.

††† $K_{OC} = K_T - K_{CAV}$.

* The tympanic membrane area of *Rhinolophus simulator*.

** Calculated from measurements of umbo motion assuming pistonlike motion with an area equal to that of the pars tensa.

*** Estimated from the change in middle ear input impedance produced by opening the middle ear cavities (Møller 1965).

**** Calculated from measurements of middle ear transmission change produced by opening the cavities (Dallos 1970) and measurements of middle ear volume (Webster and Webster 1975).

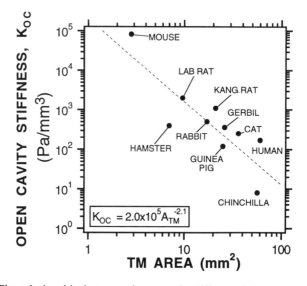

FIGURE 6.27. The relationship between the acoustic stiffness of the tympanic membrane and ossicles (the open cavity stiffness), $\mathbf{K_{OC}}$, and the area of the tympanic membrane. The units of stiffness are Pa-(mm)$^{-3}$. The dashed line is a power function fit to the data by least squares regression techniques. The probability that the exponent of the fitted function is zero is less than 1%; however, as noted in the text, the fit is heavily dependent on the isolated mouse point. The data are tabulated in Table 3.

sound pressure within the cavity, $\mathbf{P_{CAV}}$, that acts on the medial surface of the tympanic membrane such that the actual drive on the tympanic membrane is $\mathbf{P_T} - \mathbf{P_{CAV}}$. In effect, the air spaces add an impedance in series with the tympanic membrane and ossicles such that

$$\mathbf{Z_T} = \mathbf{Z_{OC}} + \mathbf{Z_{CAV}} \qquad (8)$$

where $\mathbf{Z_{OC}}$ equals the impedance of the tympanic membrane, ossicles, and cochlea and $\mathbf{Z_{CAV}}$ is the impedance of the middle ear air spaces.

The contribution of the middle ear air spaces to the middle ear input impedance is best understood at low frequencies where the impedance of the middle ear is springlike and can be defined in terms of a middle ear stiffness, K_T (Eq. 7). In the same low-frequency range, the impedance of the middle ear air spaces is determined by the stiffness of the enclosed air $\mathbf{K_{CAV}}$ and is also springlike,

$$\mathbf{Z_{CAV}} \approx \frac{-j\rho_0 c^2}{2\pi f \, \mathrm{Vol}} = \frac{-jK_{CAV}}{2\pi f} \qquad (9)$$

where ρ_0 is the density of air in standard atmospheric conditions, c is the propagation velocity of sound in air, and Vol is the middle ear air volume.

The acoustic stiffnesses of the air spaces in 10 mammalian species are noted in Table 6.3, and the relationship between total stiffness and the air volume of the middle ear is displayed in Figure 6.26. The dashed line in this figure illustrates the stiffness of the air in a closed cavity of varied volume. Clearly, much, but not all, of the interspecific variability in middle ear stiffness can be explained by differences in middle ear volume. There are species in which the total middle ear stiffness is almost completely explained by the stiffness of the middle ear air (hamster, guinea pig, and chinchilla) and there are other species where the stiffness of middle ear air greatly underestimates the total stiffness of the middle ear (mouse, kangaroo rat, and human).

It has been suggested that the middle ear air spaces also add some viscous damping to the middle ear input impedance (Onchi 1961; Zwislocki 1962). This suggestion stems from observations of the human middle ear air space that includes many small air spaces within the mastoid bone. Air moving into and out of these spaces would be subjected to viscous drag at the walls. Actual measurements of umbo motion in human cadaver middle ears show little change after widely opening the middle ear (McElveen et al. 1982; Gyo, Aritomo, and Goode 1987), and measurements of the impedance of the middle ear spaces in cats (Lynch 1981; Lynch, Peake, and Rosowski 1993) and gerbils (Ravicz, Rosowski, and Voigt 1992) reveal little air space–related damping except at very low frequencies (f < 0.1 kHz).

The role of the septa that incompletely separate the middle ear air spaces in a number of mammals (e.g., cats, chinchillas, humans, and gerbils) into several air-filled cavities is not understood. Measurements of the impedance of middle ears point out that septal foramina act to couple the entire air space together at low frequencies such that the low-frequency cavity impedance is determined by the stiffness of air within the entire middle ear. At higher frequencies, the relative acoustic impedance of these foramina increases such that they close off accessory air spaces and effectively reduce the volume of the air space behind the tympanic membrane (Onchi 1961; Zwislocki 1962; Mundie 1963; Møller 1965; Lynch 1981; Peake, Rosowski, and Lynch 1992). In the middle frequencies, Helmholtz resonances occur between the masslike impedances of the foramina and the springlike impedances of the separate air spaces. (In the cat, the Helmholtz resonance occurs at 4 kHz [Møller 1965; Guinan and Peake 1967; Lynch 1981; Peake, Rosowski, and Lynch 1992].) These resonances can have profound effects on middle ear function but only in narrow frequency ranges (Figs. 6.16, 6.17 and 6.20). The functional consequences of such "frequency-dependent" middle ear impedances and volumes are not completely understood. Several authors (e.g., Henson 1974) have noted that animals that are most sensitive to high-frequency sound tend to have small middle ear air spaces while animals that hear low-frequency sounds tend to have large air spaces. These observations led Henson (1974) to suggest that middle ear cavity septa and accompanying foramina help some animals achieve broadband auditory sensitivity by altering the effective air volume behind the eardrum, making the effective volume large at low frequencies and small at high frequencies. This suggestion requires testing.

Monotremes and a few insectivores, chiropterans, and marsupials have middle ear air spaces that are partially covered by cartilage or connective tissue rather than bone (Segall 1969; Henson 1974; Novacek 1977). The significance of these nonbony walls is not clear. From the standpoint of air-borne sound, cartilage and thick connective tissue may be expected to be as rigid as bone; however, middle ear walls made of cartilage and connective tissue may be deformed by chewing motions and other head movements. Such low-frequency distortions of the middle ear walls could introduce large low-frequency sound pressures within the middle ear air spaces that would then be coupled to the inner ear via the tympanic membrane and ossicles. It is noteworthy that the few species with nonbony middle ear walls that have been tested showed poor sensitivity to low-frequency sound (opossum *Didelphis virginianus*, Masterton, Heffner, and Ravizza 1969; echidna, Aitkin and Johnstone 1972; platypus *Ornithorhynchus anatinus*, Gates et al. 1974).

4.3.4 Stiffness of the Tympanic Membrane–Ossicular System

The data of Figure 6.26 indicate that middle ear stiffness in many ears cannot be completely explained by the stiffness of the middle ear air and suggest that the tympanic membrane and ossicles can also add a considerable stiffness. It has already been discussed in Section 4.1.3 how variations in the attachment of the anterior process of the malleus to the tympanic bone could lead to large variations in the stiffness of the ossicles, and the four stiffest ears noted in Table 6.3 (bat, mouse, echidna, and lab rat) all have firm synarthroses between the malleus and the tympanic bone (Aitkin and Johnstone 1972; Fleischer 1973). The ossicular ligaments and the tympanic membrane itself also contribute to middle ear stiffness (Zwislocki 1962; Lynch 1981; Funnell 1983; Matthews 1983). The contribution of these elements to the acoustic stiffness of the middle ear can be measured when the middle ear air spaces are widely opened ($K_{CAV} \approx 0$) or calculated by subtracting the estimated air space stiffness from the measured total stiffness (Table 6.3). Estimates of the isolated stiffness of the tympanic membrane, ossicles, and cochlea, K_{OC}, also vary regularly with the area of the tympanic membrane (Fig. 6.27), where the power function fit to the data has an exponent of approximately -2. The fit is quite significant. The probability that the exponent is 0 is less than 1%, but the relationship is heavily dependent on the isolated point for the mouse. The exponent of the power function fit to the other nine points is -1.3 and is only significant at the 5% level. The exponent fit to the nine points also is not significantly different from -1, the expected exponent if the specific acoustic stiffness of the middle ear was constant across species (Ravicz, Rosowski, and Voigt 1992). The variability in the relationship between K_{OC} and tympanic membrane area and the small number of species in which these data are

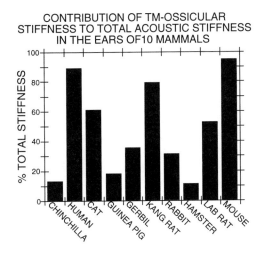

FIGURE 6.28. The relative contribution of the tympanic membrane and ossicles to the total middle ear stiffness. The species are organized along the abscissa from (left to right) lower total stiffness to greater total stiffness. The data are tabulated in Table 3.

available do not permit a clear understanding of how the stiffness of the tympanic membrane and ossicles varies with ear size.

4.3.5 Total Stiffness of the Middle Ear

As has been seen in Section 4.3.2–4.3.4, the magnitudes of the total middle ear and the stiffnesses of its components—the stiffness of the tympanic membrane, ossicles, and cochlea K_{OC}, and the stiffness of middle ear air, K_{CAV}—vary greatly among mammals (Figs. 6.25 and 6.27, Table 6.3). There is also a wide variation in the relative contribution of the two stiffness components (Fig. 6.28). Small mammals tend to have stiffer ears, but there is no clear relationship between the total stiffness and the stiffness of any one component. In the mouse, kangaroo rat, and human, the tympanic membrane and ossicles play the major role in determining total stiffness; in the chinchilla, guinea pig, and hamster, the stiffness of middle ear air dominates the total stiffness. Clearly, hypotheses that base estimates of middle ear stiffness solely on middle ear air volume (e.g., Webster 1965; Zwillenberg, Konkle, and Saunders 1981) are only valid for select species.

4.3.6 Mass of the Tympanic Membrane–Ossicular System

Although it is commonly held that the major limitation on the high-frequency performance of the middle ear is its mass (Møller 1965; Fleischer 1978; Shaw and Stinson 1983; Margolis and Shanks 1985), there is little evidence to support this conclusion. Measurements of middle ear input impedance at high frequencies do not indicate mass control (Lynch, Peake, and

Rosowski 1994), and direct measurements of the influence of additional mass on the stapes (Johnstone and Taylor 1971) and tympanic membrane (Lynch 1981) show little effect.

4.3.7 Modes of Ossicular Motion

In discussing the transduction of umbo velocity to stapes velocity, the focus was on either simple ossicular translation or rotation around a fixed incudo-malleolar axis, where evidence for both motions has been obtained in cat and human ears (Gyo, Aritomo, and Goode 1987; Decraemer, Khanna, and Funnell 1989; Donahue, Rosowski, and Peake 1991). Other authors have hypothesized more complex ossicular motions. Decraemer and Khanna (1992) have data that suggest that the cat malleus moves in three dimensions in response to sound. Hüttenbrink (1988) suggested that very large sound pressures produce an up-and-down motion of the human incus (rather than in-and-out), and Békésy (1960) and others have described a rocking and twisting of the human stapes at high sound levels (see Wever and Lawrence 1954 for a historical review of this subject).

Fleischer (1978) proposed even more complex ossicular motions for the ears of small mammals with a synarthrosis between the tympanic bone and anterior process of the malleus. In this view, the complex geometry of the tympanic-mallear attachment allows two modes of ossicular rotation. At low frequencies, the ossicles would rotate around the traditional ossicular axis drawn between the posterior ligament of the incus joint and the anterior process of the malleus. At high frequencies (f > 20 kHz), Fleischer predicted the ossicles would rotate around a second axis—perpendicular to the low-frequency axis—that is defined by the lamina that connects the manubrium to the anterior process. Measurements of ossicular motion in ears with mallear-tympanic bone synarthroses are inconclusive. Comparisons of umbo and stapes motions in the house mouse (Saunders and Summers 1982) are roughly consistent with Fleischer's hypothesis while similar measurements in the horseshoe bat (Wilson and Bruns 1983) are not.

4.3.8 The *Pars Flaccida*, an Extratympanic Route for Sound

While the concentration has been on the role of the *pars tensa* of the tympanic membrane in middle ear function, there are other routes for sound to enter the middle ear air space, notably the *pars flaccida*. Kohllöffel (1984) has suggested that the *pars flaccida* could serve to reduce the sensitivity of the middle ear to low-frequency sound. In ears with large *pars flaccidae*, the sound pressure within the middle ear air space reflects the volume velocity of both the *pars flaccida* and *pars tensa*. If the impedance of the *flaccida* is much less than the impedances of the *tensa* and the middle ear cavity, the pressure within the middle ear space would approximate the pressure in the ear canal, reducing the force that drives the middle ear, and the *tensa* would not move

(Kohllöffel 1984). However, Unge, Bagger-Sjöbäck, and Borg (1991) have observed that stiffening the *pars flaccida* has little effect on the middle ear input impedance in gerbils. Vrettakos, Dear, and Saunders (1988) and others have suggested that the *pars flaccida* may play a protective role in reducing static pressure differences across the tympanic membrane.

4.4 Possible Role for the Middle Ear Nonlinearity at High Sound Levels

It is well known that very large stimuli produce nonlinear responses of the middle ear. Demonstrations of this phenomena include altered level dependence of the response to high-intensity sounds (Guinan and Peake 1967; Buunen and Vlaming 1981) as well as static pressure-induced alterations in the response to sounds of more normal levels (Møller 1965). (The nonlinear effect of static pressure on the middle ear is the basis for clinical tympanometry, see, e.g., Lidén, Peterson, and Björkman 1970.) The location of this nonlinearity is not clear. Hüttenbrink (1988) suggested that the incudomalleolar joint has a major nonlinear role in the human ear. There is also clear evidence that the impedance of the cat annular ligament is nonlinear in response to large stapes motions (Lynch, Nedzelnitsky, and Peake 1982). The middle ear nonlinearities described so far act to reduce the ossicular motions produced by large ear canal pressures and therefore may serve to protect the ear from either very loud sounds or large static pressures (Price 1974; Hüttenbrink 1988; Price and Kalb 1991).

4.5 Nonossicular Sound Conduction to the Cochlea

The discussion of middle ear function has been limited to sound transmission via the ossicles because the ossicular route of sound conduction is clearly dominant, at least in man and the cat (Wever and Lawrence 1954; Peake, Rosowski, and Lynch 1992). Indeed, the models of Figures 6.3 and 6.4 assume that ossicular conduction is the *only* path for sound into the inner ear. However, pathological ossicular disruptions reduce, but do not eliminate, hearing sensitivity (Peake, Rosowski, and Lynch 1992). While sound can also enter the ear via sound-induced vibration of the skull and bony cochlea (e.g., bone conduction; Tonndorf 1972), otologists have long considered the residual hearing in these cases to result from direct acoustic stimulation of the oval and round windows and have devised surgical reconstructive techniques that attempt to optimize the pressure difference between the two windows (Wullstein 1956; Gaudin 1968; Goodhill 1979). The contribution of direct acoustic stimulation of the cochlear windows has been estimated by a modified middle ear model (Peake, Rosowski, and Lynch 1992). This analysis suggests that acoustic stimulation of the windows explains much, but not all, of the residual hearing after ossicular disarticulation.

4.6 Summary of Middle Ear Structure and Function

The mammalian middle ear is a complicated structure with great interspecific diversity of form and function. Although there are some basic ideas regarding the relationship of middle ear structure and function, more basic data are needed to test and generalize these conceptions. Also, the wide variations in form and the interdependence of the structures of the mammalian middle ear makes it difficult to separate the consequences of single structural features on middle ear function, and the same anatomical "adaptation" may have different functional consequences in different species. Consider the hypertrophied middle ear air spaces of the kangaroo rat and chinchilla. The impedance of the large middle ear air space in the kangaroo rat is very small compared to the impedance of the rest of the middle ear and the air space *contributes little* to middle ear function in that species. The larger middle ear air space of the chinchilla leads to an even smaller Z_{CAV}, but because the impedance of the rest of the chinchilla middle ear is smaller still, the cavity impedance is the *primary determinant* of middle ear stiffness in that animal (Table 6.3, Fig. 6.28). Therefore, before the role of the cavity in low-frequency hearing can be understood, the contribution of other middle ear structures needs to be understood.

Figure 6.29 displays an attempt to sort out the dependence of low-frequency middle ear function on the different stiffness components. The ordinate of Figure 6.29 is a measure of low-frequency middle ear gain, G_{LF}, relative to the gain expected if the middle ear air spaces were of infinite expanse ($Z_{CAV} = 0$), such that

$$G_{LF} = \frac{1}{(1 + K_{CAV}/K_{OC})}. \tag{10}$$

The abscissa of Figure 6.29 is K_{OC}/K_{CAV}, the ratio of the tympanic membrane –ossicular stiffness (measured with widely opened middle ear air spaces) to the stiffness of the intact air space. The low-frequency gain, as defined, is a strict function of the stiffness ratio and the line illustrates all possible combinations of these two values. When the middle ear air space is infinite or opened, $K_{CAV} = 0$ and the G_{LF} equals 1. When $K_{CAV} \gg K_{OC}$, the gain is less than 1 and proportional to K_{OC}/K_{CAV}. The data points in Figure 6.29 are calculated from the stiffnesses in Table 6.3.

Changes in the relative value of the middle ear stiffnesses affect the low-frequency middle ear gain in a stereotypical manner. For example, if K_{OC} is held constant, large *in*creases in the middle ear volume of the human, mouse, or kangaroo rat cause large *de*creases in the stiffness of the air spaces but only small *in*creases in low-frequency gain. On the other hand, any small *in*creases in the volume of the middle ear cavity of hamsters will lead to proportionate *de*creases in K_{CAV}, proportionate *in*creases in the stiffness ratio, and proportionate *in*creases in middle ear gain.

Analyses like that shown in Figure 6.29 can illustrate the acoustic costs

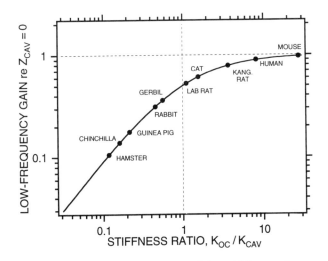

FIGURE 6.29. The interaction of the stiffnesses of the middle ear air spaces and tympanic membrane and ossicles in determining low-frequency middle ear performance, where low-frequency gain is determined by $(1 + K_{CAV}/K_{OC})^{-1}$. The solid line is the asymptotic function fit to Equation 10. The points are the calculated low-frequency gain for 10 species, and their position on the line is determined by Equation 10. The dashed vertical line marks the location where the stiffness of the air spaces and the stiffness of the TM and ossicles are equal. The dashed horizontal line marks the location where the middle ear gain is maximum and equal to 1 (after Peake unpublished).

and benefits of alterations in certain structures on middle ear sound transmission. The results of these analyses can then be weighed against other biological or ecological restrictions that act to limit the range of possible structural features, e.g., it may be beneficial from an acoustical standpoint for the hamster to increase the volume of its middle ear air spaces, but the costs to other vital functions may be too high. More relationships describing the effect of particular middle ear structures on function need to be worked out before it is completely understood how variations in ear structure affect hearing and lifestyle. This can only be done by increasing both the knowledge of the different structures and functions in mammalian middle ear and the understanding of the acoustical, mechanical, and ethological consequences of variations in structure.

5. Middle Ear Muscles

Most mammalian middle ears contain two middle ear muscles, the stapedius muscle that is attached to the stapes and the tensor tympani that attaches to the malleus, but some fossorial species have only a single muscle (Henson

1974). The relative sizes of the muscles and ossicles vary among species, with microchiropterans displaying the greatest muscle-to-ossicle size ratio (Henson 1970, 1974; Pye and Hinchcliffe 1976). The functions of the muscles have been well reviewed (Silman 1984) and include: (1) protection of the ear from loud external sounds via the acoustic reflex (Borg and Nilsson 1984), (2) protection from self-generated sounds (Carmel and Starr 1963; Henson 1970), and (3) the selective attenuation of low-frequency masking noises (Borg and Zakrisson 1975).

The acoustic reflex describes the sound-induced binaural contraction of middle ear muscles. These contractions lead to an increase in the middle ear input impedance and an attenuation of middle ear sound transmission (Fig. 6.30). The effects of muscle contraction are consistent with an increase in the stiffness of the annular ligament that supports the stapes (Rabinowitz 1977; Pang and Peake 1985, 1986). Such reflex-induced increases in stapes stiffness would directly decrease stapes motion and middle ear transmission, but only

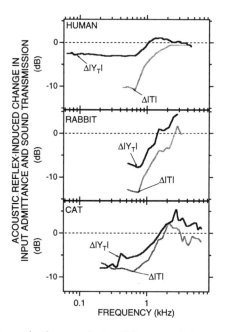

FIGURE 6.30. The change in the magnitude of the acoustic input admittance, Y_T, and sound transmission, **T**, produced by the acoustic reflex. Admittance is the inverse of impedance. Negative dB values of either $|\Delta Y_T|$ or $|\Delta T|$ indicate a decrease in middle ear motion. The reflex was elicited by contralateral sound. The human data are from Rabinowitz (1977), where ΔT was estimated from changes in loudness. The cat and rabbit data are from Møller (1965), where ΔT is defined by measurements of stapes motion. The dashed lines at 0 dB represent no change in transmission or admittance. (After Rabinowitz 1977.)

indirectly increase middle ear input impedance (decrease admittance) by changing the ossicular load on the tympanic membrane (Rabinowitz 1977; Pang and Peake 1986). Therefore, the reflex-induced change in admittance magnitude, $\Delta|\mathbf{Y}_T|$, need not be proportional to the change in transmission, $\Delta|\mathbf{T}|$ (Fig. 6.30).

Because the acoustic reflex changes annular ligament stiffness, it has its largest effect at frequencies where the middle ear is normally stiffness controlled. In man, the cat, and the rabbit, the acoustic reflex has its largest effect at frequencies below 1 kHz (Fig. 6.30). In mammals with stiffer middle ears (e.g., bats; Figs. 6.17 and 6.20), middle ear muscle contraction can effectively attenuate middle ear sound transmission at frequencies as high as 80 kHz (Henson 1970).

6. Middle and External Ear Function in Marine Mammals

The middle ear of terrestrial mammals works by transforming the sound power within airborne sound to sound power within the cochlea. The basic impediment in this process is the large difference in the specific acoustic impedances of air and cochlear fluid, which is overcome by way of the transformer mechanisms of the middle ear. When cetaceans (whales), pinnipeds (seals, sea lions, and walruses), and sirenians (manatees) moved back into the water, they were surrounded by a medium with a specific acoustic impedance more like cochlear fluid, and the transformer properties of the middle ear became less important. Indeed, several workers have suggested that the external ear is no longer operable under water and, instead, sound reaches the inner ear via bulk motion of the head or by transmission through the body walls (Norris 1964; McCormick, et al. 1970, 1980). These two proposed mechanisms present several new problems:

(1) The mammalian middle ear delivers the power that it collects to one cochlear window, the oval window. The resultant difference in stimulus pressure between the oval and round windows sets cochlear fluid into motion and stimulates the cochlear partition. Identical sound pressures delivered to the two windows result in a degraded cochlear response (Wever and Lawrence 1954; Békésy 1960; Peake, Rosowski, and Lynch 1992). In ears in which the stimulus arrives at the inner ear by nonossicular pathways, differences in the stimulus at the two windows and the resultant motion of the cochlear fluids could be small.

(2) Sound receptive mechanisms based on bulk motion of the head in water stimulate both ears simultaneously, thereby eliminating directional cues.

(3) The wavelengths of sound in water for any one frequency are more than four times longer than the wavelengths in air, and directional cues are reduced. This difference in the sound wavelength can be overcome by

increases in interaural distances and improved sensitivity to high-frequency sound, as has already occurred in many cetaceans (Reysenbach de Haan 1958).

How marine mammals deal with these problems has been a point of much discussion (Reysenbach de Haan 1958; Fraser and Purves 1960; Norris 1964; McCormick et al. 1970, 1980; Lipatov and Sointseva 1974; Norris and Harvey 1974; Ketten 1984, 1992; Popov and Supin 1990; Brill and Harder 1991).

6.1 The Anatomy of External and Middle Ears

Marine mammals can be split into two groups: the semi-aquatic pinnipeds and the completely aquatic cetaceans and sirenians.

6.1.1 The Pinniped Ear

The structures of the external and middle ears of pinnipeds have been previously reviewed (Møhl 1968; Ramprashad, Corey, and Ronald 1972; Repenning 1972). The pinna flange is absent in all but the otarids (sea lions) in which it is reduced in size. The ear canal is long and lined with a cavernous mucosal epithelia that can expand and contract in volume depending on blood supply (Møhl 1968; Repenning 1972). This distensible lining is thought to aid in closing the ear canal and maintaining ear canal pressure when the animal dives. The ossicles of pinnipeds vary in configuration; the ossicles of otarids are similar in shape to those of terrestrial carnivores, while the ossicles of the phocids (true seals) are more distinctive in form and are more massive (Doran 1879; Fleischer 1973). Repenning (1972) has suggested that phocids have small tympanic membranes, but the membrane areas are similar to those of terrestrial mammals of similar size (Fig. 6.15). The tympanic membrane to footplate area ratio appears slightly smaller in phocids (Repenning 1972) but, in general, is similar to that observed in terrestrial mammals. Pinniped middle ear air spaces are lined with cavernous mucosa that is thought to help maintain middle ear pressure when diving (Møhl 1968; Ramprashad, Corey, and Ronald 1972; Repenning 1972; Fleischer 1978). The area of the pinniped oval window is about one-third the size of the round window (these areas are roughly equal in terrestrial vertebrates) and the round window may be isolated from the rest of the middle ear air space by the cavernous mucosa and a bony round window fossula (Ramprashad, Corey, and Ronald 1972; Repenning 1972). Finally, following a recurring theme in marine mammals, the tympanic and periotic bones are relatively isolated from the skull (Repenning 1972; Fleischer 1978). The isolation of the ears is thought to allow the two ears to respond separately to sound stimuli, thereby enabling the binaural processing of directional information (Møhl 1968).

6.1.2 The Cetacean Ear

The structures of the external and middle ears of cetaceans have been reviewed by Reysenbach de Haan (1958), Purves (1966), McCormick et al. (1970), Sointseva (1973, 1990), Fleischer (1978), Ketten (1984, 1992), and Oelschläger (1986, 1990). There is no pinna in any cetacean species and the opening of the external meatus is either nonexistent or filled with wax. In some cetaceans, a few ribbons of connective tissue are the only remnants of the lateral one-third of the ear canal (Fraser and Purves 1960; McCommick et al. 1970). Closer to the tympanic membrane, the ear canal widens and may even be filled with air (Sointseva 1973, 1990). While the pinniped middle ear shares many similarities with those of terrestrial ears, the cetacean middle ear is highly derived. There are large differences in the tympanic membranes and ossicles of mysticetes (baleen whales) and odontocetes (toothed whales), but in both groups:

(1) The ossicles are massive and of unusual shape (Doran 1879; Fleischer 1978).
(2) The malleus is firmly attached to the tympanic bone—by bone in odontocetes and cartilage in mysticetes (Reysenbach de Haan 1958; McCormick et al. 1970; Sointseva 1990).
(3) The tympanic membrane is nearly unrecognizable and its attachment to the ossicles is tenuous (McCormick et al. 1970; Fleischer 1978).
(4) As in pinnipeds, the middle ear space is lined with a cavernous mucosa.
(5) The tympanic-periotic complex is almost completely isolated from the rest of the skull (Reysenbach de Hann 1959; Fleischer 1978).

The differences in the middle ear structures of mysticetes and odontocetes may be related to differences in the frequency range of hearing in these animals (Reysenbach de Haan 1959; Fleischer 1978; Ketten 1984, 1992). The sirenian middle ear has features in common with the mysticete ear and both have been suggested to be most efficient in transmitting low-frequency sounds (Ketten 1992, personal communication).

6.2 External and Middle Ear Functions in Water

6.2.1 Sound in Water

Water has much different acoustic properties than air; it is about 1000 times denser and nearly incompressible. The speed of sound in water is more than four times that of air, and the characteristic acoustic impedance of water is nearly 4000 times that of air. Soft body tissues (e.g. muscle and fat) have acoustic properties approximating that of water, while dense bone may have an impedance that is two to four times larger (Norris 1964; McCormick et al. 1970). The close match of impedance in water and flesh allows sound energy to flow from water to body tissue with little loss.

6.2.2 Production of Cochlear Fluid Motion by Waterborne Sound

The effective stimulus to the inner ears of mammals (and vertebrates in general) is the motion of hair cell stereocilia with respect to the body of the hair cell (Dallos 1981). These stereocilia motions are produced by the motion of the cochlear partition, which in turn is produced by the motion of the inner ear fluids. As discussed in this section, cochlear fluid motion is produced by the differential stimulation of the cochlear windows (Wever and Lawrence 1954; McCormick et al. 1980). The function of the external and middle ears in air is to gather sound power and deliver it preferentially to the oval window, thereby producing a differential stimulus. In water, sound power may actually flow better through the surrounding tissues of the head rather than through the external and middle ears. However, while the middle ear may have no advantage for the conduction of waterborne sound power to the cochlea, it may serve an important role in allowing cochlear fluid motion under water (McCormick et al. 1980).

Waterborne sound acts on the ear by vibrating and compressing the head and bony inner ear. This stimulus mimics the direct vibratory and compressional stimulation applied to the skull during bone conduction audiometry. The mechanism of bone conduction has been studied extensively (e.g., Wever and Lawrence 1954; Békésy 1960; Tonndorf 1972), and several stimulus pathways to the inner ear have been proposed:

(1) Bulk motion of the head can cause relative motion between the ossicles, cochlear fluids, and cochlea, mimicking air-conducted sound.
(2) Compression of the bone surrounding the external and middle ear air spaces produces airborne sound pressures that act on the tympanic membrane and cause ossicular motion.
(3) The nonrigid stapes and round window membrane allow cochlear fluid motion when the cochlear walls are compressed by sound.

The middle ear plays an important role in all of these mechanisms, and alterations of the middle ear can greatly reduce the behavioral response to bone conduction stimuli (Tonndorf and Tabor 1962; Tonndorf 1972). Similar mechanisms have been proposed to explain hearing in marine mammals (McCormick et al. 1970, 1980; Lipatov and Sointseva 1974).

McCormick and colleagues (1970) demonstrated that, while removing the malleus had little effect on the cochlear response to waterborne sound in a dolphin, removing the stapes caused a large decrease in cochlear sensitivity. They interpreted this result as an indication that waterborne sound produced relative motion of the stapes and cochlea that then produced cochlear fluid motion. Lipatov and Sointseva (1974, cited in Sointseva 1990) suggested that sound transmitted through the soft tissue surrounding the ear can compress air within the ear canal and middle ear, generating an acoustic sound pressure that then stimulates the middle ear.

Several authors have also suggested specializations that would channel tissue-conducted sound to the inner ear. Norris (1964) suggested that a specialization of the dolphin jaw and skull would serve to gather waterborne sound energy and conduct it to the inner ear. This mechanism has been supported by acoustic measurements of sound conduction in dolphin heads (Norris and Harvey 1974) and behavioral measurements made after shielding the jaw from sound (Brill and Harder 1991). Popov and Supin (1990), on the other hand, have measurements that suggest that the dolphin ear is most sensitive to sounds directed at the external canal.

Unfortunately, there are little data describing the function of the external and middle ears in marine mammals. Earlier suggestions that these ears worked like terrestrial ears, with sound being conducted down the ear canal to the tympanic membrane (Fraser and Purves 1960), are probably incorrect. Instead, it seems likely that sound conducted through the body tissues plays a large role in how marine mammals hear under water. The fact that the ears of the semi-aquatic pinnipeds appear more sensitive to waterborne sound (Møhl 1968; Moore and Schusterman 1987) argues that these animals utilize two systems for sound conduction: the terrestrial external and middle ears for airborne sound and some form of bone conduction mechanism for waterborne sound that depends on the middle ear.

7. Summary

The external and middle ears of mammals act by selectively transferring acoustical power of certain frequencies from the environment to the inner ear. In performing this function, the auditory periphery is constrained by the size and structure of its varying components. Although few specifics are understood, ears that seem to act best at transferring high-frequency sound energy are small in dimension and stiff. Ears that act best at transferring low-frequency sounds are larger in dimension and less stiff. Some ears are capable of acting in both frequency ranges and seem to contain specializations from both groups, e.g., the malleus of the domestic cat is neither freestanding nor tightly coupled to the tympanic bone but, instead, is connected to the tympanic bone by a small anterior process (Fleischer 1978). Similarly, the foramen that couples the two large middle ear air spaces in the cat enables a large middle ear air volume at low frequencies and a much smaller effective volume at high frequencies.

This chapter has suggested several relationships between form and function that need further investigation.

(1) The interaction of mammalian body size and hearing capability has been well documented by Masterton, Heffner, and Ravizza (1969) and Heffner and Heffner (1992). However, the interaction between ear size and auditory capabilities is only now becoming clear. These newer findings raise old questions in new lights. Is it true that mammals developed high-

frequency hearing in order to compensate for the small size of their head in auditory localization tasks or were the small ears of small mammals by necessity restricted to high audio frequencies? In the limited number of ears in which functional measurements have been made, there does seem to be an inverse relationship between middle ear dimensions and stiffness (Figs. 6.25 and 6.26), where increased stiffness can be loosely related with increased low-frequency thresholds (Rosowski 1992). Does this relationship hold for all mammals? Measurements of function and structure in a wider variety of mammals are needed before this question can be answered. Of particular interest would be measurements in animals with extraordinary high- and low-frequency hearing, as well as measurements on species with widely varying ossicular and middle ear cavity structures.

(2) The total stiffness of the middle ear is determined by the combination of the stiffness of the middle ear air, K_{CAV}, and the stiffness of the tympanic membrane, ossicles, and cochlea, K_{OC}. There is no simple rule that relates the magnitudes of these two stiffness components (Figs. 6.25, 6.26, 6.27, and 6.28). There are low-stiffness ears where either K_{CAV} dominates (e.g., the chinchilla) or K_{OC} dominates (e.g., humans), and high-stiffness ears may also be dominated by either component. Questions of interest include: Are ears in which the stiffness is dominated by either component more efficient at gathering sounds of some frequency range? Is there some interaction between external ear function and the stiffness of either component? Are there nonauditory reasons, e.g., the persistence of a large stapedial artery within the middle ear or the lack of bony middle ear walls, which significantly influence the stiffness of either component? Answers to these questions require more measurements of middle and external ear structures and functions in a wide variety of mammals, including measurements of the role of the middle ear cavity in sound transmission and estimates of power utilization. A better understanding of the impact of middle ear cavity size on nonauditory function is also required.

(3) Two approaches to understanding the role of the external and middle ear in limiting auditory function have been described. Comparisons of the power collected by the external and middle ear with auditory thresholds in a few species (Shaw and Stinson 1983; Rosowski et al. 1986; Rosowski 1991) suggest that the shape of the auditory function is completely determined by external and middle ear function. Comparisons of this sort are needed in a wide variety of species with different auditory capabilities, e.g., the present test population excludes small mammals with good high-frequency hearing. The second approach concerns comparisons of middle ear structure with behavioral measurements of hearing (Rosowski and Graybeal 1990; Rosowski 1992). These comparisons were made on a limited number of mammalian species with both known middle ear dimensions and behavioral audiograms. Better measurements of middle ear structure in more mammals with known audiograms will help refine the understanding of auditory structure and function.

Acknowledgments. This work was supported by a grant from the National Institute for Deafness and Communicative Disorders (DC-R01-00194). Thanks go to Bill Peake, Mike Ravicz, and Sunil Puria for many helpful discussions, and extensive comments and suggestions on drafts of this work, and Michael L. Dickens for programming the ear canal model used in generating Figure 6.10.

References

Aitkin LM, Johnstone BM (1972) Middle ear function in a monotreme: The echidna (*Tachyglossus aculeatus*). J Exp Zool 180:245–250.

Allen J (1986) Measurements of eardrum acoustic impedance. In: Allen JB, Hall JL, Hubbard A, Neely ST, Tubis A (eds) Peripheral Auditory Mechanisms. New York: Springer-Verlag, pp. 44–51.

Allin EF (1975) Evolution of the mammalian middle ear. J Morphol 147:403–437.

Allin EF (1986) The auditory apparatus of advanced mammal-like reptiles and early mammals. In: Hotton N III, Maclean PD, Roth JJ, Roth EC (eds) The Ecology and Biology of Mammal-like Reptiles. Washington, DC: Smithsonian Institution Press, pp. 283–294.

Allin EF, Hopson JA (1992) Evolution of the auditory system in synapsida ("mammal-like" reptiles and primitive mammals) as seen in the fossil record. In: Webster DB, Fay RR, Popper AN (eds) The Evolutionary Biology of Hearing. New York: Springer-Verlag, pp. 587–614.

Aritomo H, Goode RL (1988) Cochlear input impedance in fresh human temporal bones. Otolaryngol Head Neck Surg 97:136.

Bast TH, Anson BJ (1949) The Temporal Bone and the Ear. Springfield, IL: Charles C. Thomas.

Békésy G von (1960) Experiments in Hearing. New York: McGraw-Hill.

Beranek LL (1949) Acoustic Measurements. New York: John Wiley and Sons.

Beranek LL (1954) Acoustics. New York: McGraw-Hill.

Blauert J (1983) Spatial Hearing. Cambridge, MA: MIT Press.

Borg E (1968) A quantitative study of the effect of the acoustic stapedius reflex on sound transmission through the middle ear of man. Acta Otolaryngol 66:461–472.

Borg E, Nilsson R (1984) Acoustic reflex in industrial noise. In: Silman S (ed) The Acoustic Reflex. New York: Academic Press, pp. 413–440.

Borg E, Zakrisson JE (1974) Stapedius muscle and monaural masking. Acta Otolaryngol 94:385–393.

Brill RL, Harder PJ (1991) The effects of attenuating returning echolocation signals at the lower jaw of a dolphin (*Tursiops truncatus*). J Acoust Soc Am 89:2851–2857.

Buss IO, Estes JA (1971) The functional significance of movements and positions of the pinnae of African elephants *Loxodonta africana*. J Mammal 52:21–27.

Butler RA, Belendiuk K (1977) Spectral cues utilized in the localization of sound in the median sagittal plane. J Acoust Soc Am 61:1264–1269.

Buunen TJF, Vlaming MSMG (1981) Laser-Doppler velocity meter applied to tympanic membrane vibrations in cat. J Acoust Soc Am 69:744–750.

Calford MB, Pettigrew JD (1984) Frequency dependence of directional amplification at the cat's pinna. Hear Res 14:13–19.

Carlile S (1990a) The auditory periphery of the ferret. I: Directional response properties and the pattern of interaural level differences. J Acoust Soc Am 88:2180–2195.

Carlile S (1990b) The auditory periphery of the ferret. II: The spectral transformations of the external ear and their implications for sound localization. J Acoust Soc Am 88:2196–2204.

Carlile S, Pettigrew AG (1987) Directional properties of the auditory periphery in the guinea pig. Hear Res 31:111–122.

Carmel PW, Starr A (1963) Acoustic and nonacoustic factors modifying middle ear muscle activity in waking cats. J Neurophysiol 26:598–616.

Cockerell TDA, Miller LI, Printz M (1914) The auditory ossicles of American rodents. Bull Am Mus Nat Hist 33:347–364.

Coles RG, Guppy A (1986) Biophysical aspects of directional hearing in the Tammar wallaby Macropus eugenii. J Exp Biol 121:371–394.

Dahmann H (1929) Zur Physiologie des Hörens; experimentelle Untersuchungen über die Mechanik der Gehörknöchelchenkette sowie über deren Verhalten auf Ton und Luftdruck. Zeits f Hals-Nasen-Ohrenheilk 24:462–497.

Dallos P (1970) Low frequency auditory characteristics: Species dependence. J Acoust Soc Am 48:489–499.

Dallos P (1973) The Auditory Periphery. New York: Academic Press.

Dallos P (1981) Cochlear physiology. Ann Rev Psychol 32:153–190.

Dancer A, Franke R (1980) Intracochlear sound pressure measurements in guinea pigs. Hear Res 2:191–205.

Dear SP (1987) Impedance and sound transmission in the auditory periphery of the chinchilla. PhD Thesis, University of Pennsylvania, Philadelphia.

Décory L (1989) Origine des différences interspecifiques de susceptibilité an bruit. Thése de Doctorat de l'Université de Bordeaux, France.

Decraemer WF, Khanna SM (1992) Three dimensional displacement of the umbo in cat. Abstracts of the Fifteenth Midwinter Meeting of the Association for Research in Otolaryngology 156.

Decraemer WF, Khanna SM, Funnel WRJ (1989) Interferometric measurement of the amplitude and phase of tympanic membrane vibrations in cat. Hear Res 38:1–18.

Decraemer WF, Khanna SM, Funnell WRJ (1991) Malleus vibration mode changes with frequency. Hear Res 54:305–318.

Desoer CA, Kuh ES (1969) Basic Circuit Theory. New York: McGraw-Hill.

DiMaio FHP, Tonndorf J (1978) The terminal zone of the external auditory meatus in a variety of mammals. Arch Otolaryngol 104:570–575.

Donahue KM, Rosowski JJ, Peake WT (1991) Can the motion of the human malleus be described as pure rotation? Abstracts of the Fourteenth Midwinter Meeting of the Association for Research in Otolaryngology 52.

Doran AHG (1879) Morphology of the mammalian ossicula auditus. Trans Linn Soc 1:371–497.

Drescher DG, Eldredge DH (1974) Species differences in cochlear fatigue related to acoustics of outer and middle ears of guinea pig and chinchilla. J Acoust Soc Am 56:929–934.

Egolf DP (1980) Techniques for modeling the hearing aid receiver and associated tubing. In: Studebaker GA, Hochberg I (eds) Acoustical Factors Affecting Hearing Aid Performance. Baltimore, MD: University Park Press, pp. 297–319.

Fay RR (1988) Hearing in Vertebrates: A Psychophysical Source Book. Winnetka, IL: Hill-Fay Associates.

Fleischer G (1973) Studien am Skelett des Gehörorgans der Säugetiere einschliesslich des Menschen. Säugetierkundl Mitteilungen (München) 21:131–239.

Fleischer G (1978) Evolutionary principles of the mammalian middle ear. Adv Anat Embryol Cell Biol 55:3–69.

Fletcher NH, Thwaites S (1979) Physical models for the analysis of acoustical systems in biology. Quart Rev Biophys 12:25–65.

Fletcher NH, Thwaites S (1988) Obliquely truncated simple horns: Idealized models for vertebrate pinnae. Acoustica 65:194–204.

Fraser FC, Purvis PE (1960) Anatomy and function of the cetacean ear. Proc R Soc Br 152:62–77.

Funnell WR (1983) On the undamped natural frequencies and mode shapes of a finite-element model of the cat eardrum. J Acoust Soc Am 73:1657–1661.

Funnell WR, Laszlo CA (1977) Modeling of the cat eardrum as a thin shell using the finite-element method. J Acoust Soc Am 63:1461–1467.

Funnell WR, Laszlo CA (1982) A critical review of experimental observations on eardrum structure and function. ORL 44:181–205.

Funnell WR, Decraemer WF, Khanna SM (1987) On the damped frequency response of a finite-element model of the cat eardrum. J Acoust Soc Am 81:1851–1859.

Funnell WR, Decraemer WF, Khanna SM (1992) On the degree of rigidity of the manubrium in a finite-element model of the cat eardrum. J Acoust Soc Am 91:2082–2090.

Gates GR, Saunders JC, Bock GR, Aitkin LM, Elliott MA (1974) Peripheral auditory function in the platypus *Ornithorhynchus anatinus*. J Acoust Soc Am 56:152–156.

Gaudin EP (1968) Hearing improvement by sound isolation of the round window. Arch Otolaryngol 87:376–377.

Goodhill V (1979) Ear: Diseases, Deafness and Dizziness. Hagerstown, MD: Harper & Row.

Graham MD, Reams D, Perkins R (1978) Human tympanic membrane-malleus attachment. Ann Otol Rhinol Laryngol 87:426–431.

Gray AA (1913) Notes on the comparative anatomy of the middle ear. J Anat Physiol (London) 47:391–413.

Guinan JJ Jr, Peake WT (1967) Middle ear characteristics of anesthetized cats. J Acoust Soc Am 41:1237–1261.

Gundersen T (1971) Prostheses in the Ossicular Chain. Baltimore, MD: University Park Press.

Guppy A, Coles RB (1988) Acoustical and neural aspects of hearing in the Australian gleaning bats *Macroderma gigas* and *Nyctophilus gouldi*. J Comp Physiol A 162:653–668.

Gyo K, Aritomo H, Goode RL (1987) Measurement of the ossicular vibration ratio in human temporal bones by use of a video measuring system. Acta Otolaryngol 103:87–95.

Heffner HE, Masterton RB (1980) Hearing in glires: Domestic rabbit, cotton rat, feral house mouse and kangaroo rat. J Acoust Soc Am 68:1584–1599.

Heffner RS, Heffner HE (1990) Vestigial hearing in a fossorial mammal, the pocket gopher (*Geomys bursarius*). Hear Res 46:239–252.

Heffner RS, Heffner HE (1992) Evolution of sound localization in mammals. In: Webster DB, Fay RR, Popper AN (eds) The Evolutionary Biology of Hearing. New York: Springer-Verlag, pp. 691–716.

Heffner RS, Heffner HE, Stichman N (1982) Role of the elephant pinna in sound localization. Anim Behav 30:628–629.

Helmholtz HL von (1868) Die Mechanik der Gehörknöchelchen und des Trommelfells. Pflüg Arch ges Physiol 1:1–60.

Henson OW Jr (1961) Some morphological and functional aspects of certain structures of the middle ear in bats and insectivores. Univ Kansas Sci Bull 42:151–255.

Henson OW Jr (1970) The ear and audition. In: Wimsatt WA (ed) The Biology of Bats, Volume 2. New York: Academic Press, pp. 181–262.

Henson OW Jr (1974) Comparative anatomy of the middle ear. In: Keidel WD, Neff WD (eds) The Handbook of Sensory Physiology: The Auditory System V/l. New York: Springer-Verlag, pp. 39–110.

Hill RW, Christian DP, Veghte JH (1980) Pinna temperature in exercising jackrabbits *Lepus californicus*. J Mammal 61:30–38.

Hinchcliffe R, Pye A (1969) Variations in the middle ear of the Mammalia. J Zool (London) 157:277–288.

Hudde H (1983) Measurement of the eardrum impedance of human ears. J Acoust Soc Am 73:242–247.

Hudde H, Schröter J (1980). The equalization of artificial heads without exact replication of eardrum impedance. Acoustica 44:301–307.

Hunt RM Jr (1974) The auditory bulla in carnivora: An anatomical basis for reappraisal of carnivore evolution. J Morphol 43:21–76.

Hunt RM Jr (1987) Evolution of Aeluroid carnivora: Significance of the auditory structures in the nimravid cat *Dinictus*. Am Mus Novit 2886:1–74.

Hunt RM Jr, Korth WW (1980) The auditory region of Dermoptera: Morphology and function relative to other living mammals. J Morphol 164:167–211.

Hüttenbrink KB (1988) The mechanics of the middle ear at static air pressures. Acta Otolaryngol Suppl 451:1–35.

Jen PH, Chen D (1988) Directionality of sound pressure transformation at the pinna of echolocating bats. Hear Res 34:101–118.

Johnstone BM, Taylor K (1971) Physiology of the middle ear transmission system. Otolaryngol Soc Aust 3:225–228.

Keefe DH (1984) Acoustical wave propagation in cylindrical ducts: Transmission line parameter approximations for isothermal and non-isothermal boundary conditions. J Acoust Soc Am 75:58–62.

Keen JA, Grobbelaar CS (1941) The comparative anatomy of the tympanic bulla and auditory ossicles with a note suggesting their function. Trans R Soc S Afr 28:307–329.

Kemp DT (1978) Stimulated acoustic emissions from the human auditory system. J Acoust Soc Am 64:1386–1391.

Kermack KA, Musset F (1983) The ear in mammal-like reptiles and early mammals. Acta Palaeontol Polonica 28:148–158.

Ketten DR (1984) Correlations of morphology with frequency for odontocete cochlea: Systematics and topology. PhD Thesis, The Johns Hopkins University, Baltimore, MD.

Ketten DR (1992) The marine mammal ear: Specializations for aquatic audition and echolocation. In: Webster DB, Fay RR, Popper AN (eds) The Evolutionary Biology of Hearing. New York: Springer-Verlag, pp. 717–754.

Khanna SM, Stinson MR (1985) Specification of the acoustical input to the ear at high frequencies. J Acoust Soc Am 77:577–589.

Khanna SM, Tonndorf J (1969) Middle ear power transfer. Archives Klin exp Ohren-Nasen Kehlkopfherkd Heilk 193:78–88.

Khanna SM, Tonndorf J (1972) Tympanic membrane vibrations in cats studied by time-average holography. J Acoust Soc Am 51:1904–1920.

Khanna SM, Tonndorf J (1978) Physical and physiological principles controlling auditory sensitivity in primates. In: Noback R (ed) Neurobiology of Primates. New York: Plenum Press, pp. 23–52.

Killion MC, Dallos P (1979) Impedance matching by the combined effects of the outer and middle ear. J Acoust Soc Am 66:599–602.

Kim DO (1980) Cochlear mechanics: Implications of electrophysiological and acoustical observations. Hear Res 2:297–317.

Kinsler LE, Frey AR (1962) Fundamentals of Acoustics. New York: John Wiley and Sons.

Kirikae I (1960) The Structure and Function of the Middle Ear. Tokyo: University of Tokyo Press.

Kobrak HG (1959) The Middle Ear. Chicago, IL: University of Chicago Press.

Kohllöffel LUE (1984) Notes on the comparative mechanics of hearing. III. On Shrapnell's membrane. Hear Res 13:83–88.

Kringlebotn M (1988) Network model for the human middle ear. Scand Audiol 17: 75–85.

Kringlebotn M, Gundersen T (1985) Frequency characteristics of the middle ear. J Acoust Soc Am 77:159–164.

Kuhn GF (1977) Model for the interaural time differences in the azimuthal plane. J Acoust Soc Am 62:157–167.

Kuhn GF (1979) The pressure transformation from a diffuse sound field to the external ear and to the body and head surface. J Acoust Soc Am 65:991–1000.

Kuhn GF (1987) Physical acoustics and measurements pertaining to directional hearing. In: Yost WA, Gourevitch G (eds) Directional Hearing. New York: Springer-Verlag, pp. 3–25.

Lawrence BD, Simmons JA (1982) Echolocation in bats: The external ear and perception of the vertical positions of targets. Science 218:481–483.

Lay DM (1972) The anatomy, physiology, functional significance and evolution of specialized hearing organs of Gerbilline rodents. J Morphol 138:41–120.

Legouix JP, Wisner A (1955) Role fonctionnel des bulles tympaniques géantes de certains rongeurs (Meriones). Acoustica 5:208–216.

Lidén G, Peterson JL, Björkman G (1970) Tympanometry. Arch Otolaryngol 92:248–257.

Lim DJ (1968) Tympanic membrane. Part II. Pars flaccida. Acta Otolaryngol 66:515–532.

Lin Q (1990) Speech production theory and articulatory speech synthesis. Technical Report of the Royal Institute of Technology, Stockholm. Regart 90–1, 186 pg.

Lipatov NV, Sointseva GN (1974) Morpho-functional features of the biomechanics of the middle ear of dolphins. Bionics 8:113–117 (in Russian).

Lynch TJ III (1981) Signal processing by the cat middle ear: Admittance and transmission, measurements and models. ScD Thesis, Massachusetts Institute of Technology, Cambridge.

Lynch TJ III, Nedzelnitsky V, Peake WT (1982) Input impedance of the cochlea in cat. J Acoust Soc Am 72:108–130.

Lynch TJ III, Peake WT, Rosowski JJ (1994) Measurements of the acoustic input impedance of cat ears: 10 Hz to 22 kHz. J Acoust Soc Am (in press).

Manley GA, Johnstone BM (1974) Middle ear function in the guinea pig. J Acoust Soc Am 56:571–576.

Manley GA, Irvine DR, Johnstone BM (1972) Frequency response of bat tympanic membrane. Nature 237:112–113.

Margolis RH, Shanks JE (1985) Tympanometry. In: Katz J (ed.) Handbook of Clinical Audiology. Baltimore, MD: Williams & Wilkens, pp. 438–475.

Marquet J, Van Camp KJ, Creten WL (1973) Topics in physics and middle-ear surgery. Acta Otorhinolaryngol Belgium 27:137.

Masterton RB, Heffner HE, Ravizza R (1969) The evolution of human hearing. J Acoust Soc Am 45:966–985.

Matthews JW (1980) Mechanical modeling of nonlinear phenomena observed in the peripheral auditory system. ScD Thesis, Washington University, St. Louis, MO.

Matthews JW (1983) Modeling reverse middle ear transmission of acoustic distortion signals. In: deBoer E, Viergever MA (eds) Mechanics of Hearing. Delft, The Netherlands: Delft University Press, pp. 11–18.

McCormick JG, Wever EG, Palin J, Ridgway SH (1970) Sound conduction in the dolphin ear. J Acoust Soc Am 48:1418–1428.

McCormick JG, Wever EG, Ridgway SH, Palin J (1980) Sound reception in the porpoise as it relates to echolocation. In: Busnel RG, Fish JF (eds) Animal Sonar Systems. New York: Plenum Press, pp. 449–467.

McElveen JT, Goode RL, Miller C, Falk SA (1982) Effect of mastoid cavity modification on middle ear sound transmission. Ann Otol Rhinol Laryngol 91:526–532.

Merchant SN, Ravicz ME, Rosowski JJ (1992) The acoustic input impedance of the stapes and cochlea in human temporal bones. Abstracts of the Fifteenth Midwinter Meeting of the Association for Research in Otolaryngology 98.

Metz O (1946) The acoustic impedance measured on normal and pathological ears. Acta Otolaryngol Suppl 63:1–254.

Middlebrooks JC, Makous JC, Green DM (1989) Directional sensitivity of sound-pressure levels in the human ear canal. J Acoust Soc Am 86:89–108.

Møhl B (1968) Hearing in seals. In: Harrison RJ, Hubbard RC, Peterson RS, Rice CE, Schusterman RJ (eds) Behavior and Physiology of Pinnipeds. New York: Appleton-Century-Crofts, pp. 172–195.

Møller AR (1961) Network model of the middle ear. J Acoust Soc Am 33:168–176.

Møller AR (1965) Experimental study of the acoustic impedance of the middle ear and its transmission properties. Acta Otalaryngol 60:129–149.

Møller AR (1974) The acoustic middle ear muscle reflex. In: Keidel WD, Neff WD (ed) Handbook of Sensory Physiology: Vol V/l: Auditory System. New York: Springer-Verlag, pp. 519–548.

Molvær O, Vallersnes FM, Kringlebotn M (1978) The size of the middle ear and the mastoid air cells. Acta Otolaryngol 85:24–32.

Moore PWB, Schusterman RJ (1987) Audiometric assessment of northern fur seals, Callorhinus ursinus. Mar Mammal Sci 3:31–53.

Mundie R (1963) The impedance of the ear—A variable quantity. US Army Med Res Rep 576, pp. 63–85.

Musicant AD, Chan JCK, Hind JE (1990) Direction-dependent spectral properties of cat external ear: New data and cross-species comparisons. J Acoust Soc Am 87:757–781.

Nedzelnitsky V (1974) Measurements of sound pressure in the cochleae of anesthetized cats. In: Zwicker E, Terhardt E (eds) Facts and Models of Hearing. New York: Springer-Verlag, pp. 45–53.

Nedzelnitsky V (1980) Sound pressures in the basal turn of the cat cochlea. J Acoust

Soc Am 68:1676–1689.

Norris KS (1964) Some problems of echolocation in Cetacea. In: Tavolga WN (ed) Marine Bioacoustics. New York: Pergamon Press, pp. 317–336.

Norris KS, Harvey GW (1974) Sound transmission in the porpoise head. J Acoust Soc Am 56:659–664.

Novacek MJ (1977) Aspects of the problem of variation, origin and evolution of the eutherian auditory bulla. Mammal Rev 7:131–150.

Nuttall AL (1974) Measurement of the guinea pig middle ear transfer characteristic. J Acoust Soc Am 56:1231–1238.

Oelschläger HA (1986) Comparative morphology and evolution of the otic region in toothed whales, Cetacea, Mammalia. Am J Anat 177(3):353–368.

Oelschläger HA (1990) Evolutionary morphology and acoustics in the dolphin skull. In: Thomas J, Kastelein R (eds) Sensory Abilities of Cetaceans. New York: Plenum Press, pp. 137–162.

Onchi Y (1961) Mechanism of the middle ear. J Acoust Soc Am 33:794–805.

Pang XD, Peake WT (1985) A model for changes in middle ear transmission by stapedius muscle contraction. J Acoust Soc Am 78:S13.

Pang XD, Peake WT (1986) How do contractions of the stapedius muscle alter the acoustic properties of the middle ear? In: Allen JB, Hall JL, Hubbard A, Neely ST, Tubis A (ed) Peripheral Auditory Mechanisms. New York: Springer-Verlag, pp. 36–43.

Peake WT, Guinan JJ Jr (1967) Circuit model for the cat's middle ear. MIT Quart Prog Rep 84:320–326.

Peake WT, Rosowski JJ (1991) Impedance matching optimum velocity and ideal middle ears. Hear Res 53:1–6.

Peake WT, Rosowski JJ, Lynch TJ III (1992) Middle ear transmission: Acoustic vs. ossicular coupling in cat and human. Hear Res 57:245–268.

Peterson EA, Levison M, Lovett S, Feng A, Dunn SH (1974) The relation between middle ear morphology and peripheral auditory function in rodents. I. Sciuridae. J Aud Res 14:227–242.

Phillips DP, Calford MB, Pettigrew JD, Aitkin LM, Semple MN (1982) Directionality of sound pressure transformation at the cat's pinna. Hear Res 8:13–28.

Plassmann W, Webster DB (1992) Parallel evolution of low frequency sensitivity in Old World and New World desert rodents. In: Webster DB, Fay RR, Popper AN (eds) The Evolutionary Biology of Hearing. New York: Springer-Verlag, pp. 637–654.

Pocock RI (1921) The auditory bulla and other cranial characters in the Mustelidae. Proc Zool Soc London, pp. 473–486.

Pocock RI (1928) The structure of the auditory bulla in the Procyonidae and the Ursidae with a note on the bulla of *Hyaena*. Proc Zool Soc London, pp. 963–974.

Popov V, Supin S (1990) Localization of the acoustic window at the dolphin's head. In: Thomas J, Kastelein R (eds) Sensory Abilities of Cetaceans. New York: Plenum Press, pp. 417–426.

Price GR (1974) Upper limit of stapes displacement: Implications for hearing loss. J Acoust Soc Am 56:195–197.

Price GR, Kalb JT (1991) Insight into hazard from intense impulses from a mathematical model of the ear. J Acoust Soc Am 90:219–227.

Puria S, Allen JB (1991) A parametric study of cochlear input impedance. J Acoust Soc Am 89:287–309.

Purves PE (1966) Anatomy and physiology of the outer and middle ear in cetaceans. In: Norris BK (ed) Whales, Dolphins and Porpoises. Berkeley: University of California Press, pp. 320–380.

Pye A, Hinchcliffe R (1976) The comparative anatomy of the ear. In: Hinchcliffe R, Harrison D (eds) Scientific Foundations of Otolaryngology. London: William Heineman, pp. 184–202.

Rabbitt RD, Holmes MH (1986) A fibrous dynamic continuum model of the tympanic membrane. J Acoust Soc Am 80:1716–1728.

Rabinowitz WM (1977) Acoustic reflex effects on the input admittance and transfer characteristics of the human middle ear. PhD Thesis, Massachusetts Institute of Technology, Cambridge.

Rabinowitz WM (1981) Measurement of the acoustic input admittance of the human ear. J Acoust Soc Am 70:1025–1035.

Ramprashad F, Corey S, Ronald K (1972) Anatomy of the seal's ear (*Pagophilus groenlandicus*). In: Harrison RJ (ed) Functional Anatomy of Marine Mammals. New York: Academic Press, pp. 263–306.

Ravicz ME (1990) Acoustic impedance of the gerbil ear. MS Thesis, Boston University, Boston, MA.

Ravicz ME, Rosowski JJ, Voigt HF (1992) Sound power collection by the auditory periphery of the Mongolian gerbil *Meriones unguiculatus*. I. Middle ear input impedance. J Acoust Soc Am 92:157–177.

Rayleigh JWS (1945) The Theory of Sound, Volume II. New York: Dover Publications.

Relkin EM, Saunders JC (1980) Displacement of the malleus in neonatal golden hamsters. Acta Otolaryngol 90:6–15.

Repenning CA (1972) Underwater hearing in seals: Functional morphology. In: Harrison RJ (ed) Functional Anatomy of Marine Mammals. New York: Academic Press, pp. 307–331.

Reysenbach de Haan FW (1958). Hearing in whales. Acta Otolaryngol Suppl 134: 1–114.

Rhode WS (1971) Observations of the vibration of the basilar membrane in squirrel monkeys using the Mössbauer technique. J Acoust Soc Am 49:1218–1231.

Rhode WS (1978) Some observations on cochlear mechanics. J Acoust Soc Am 64: 158–176.

Robles L, Ruggero MA, Rich NC (1986) Basilar membrane mechanics at the base of the cochlea. I. Input-output functions, tuning curves and response phases. J Acoust Soc Am 80:1364–1374.

Rosowski JJ (1991) The effects of external and middle ear filtering on auditory threshold and noise-induced hearing loss. J Acoust Soc Am 90:124–135.

Rosowski JJ (1992) Hearing in transitional mammals: Predictions from the middle ear anatomy and hearing capabilities of extant mammals. In: Webster DB, Fay RR, Popper AN (eds) The Evolutionary Biology of Hearing. New York: Springer-Verlag, pp. 615–632.

Rosowski JJ, Graybeal A (1991) What did *Morganucodon* hear? Zool J Linn Soc 101:131–168.

Rosowski JJ, Peake WT, Lynch TJ III (1984) Acoustic input admittance of the alligator lizard ear: Nonlinear features. Hear Res 16:205–223.

Rosowski JJ, Carney LH, Lynch TJ III, Peake WT (1986) The effectiveness of the external and middle ears in coupling acoustic power into the cochlea. In: Allen JB, Hall JL, Hubbard A, Neely ST, Tubis A (eds) Peripheral Auditory Mechanisms.

New York: Springer-Verlag, pp. 3–12.

Rosowski JJ, Carney LH, Peake WT (1988) The radiation impedance of the external ear of cat: Measurements and applications. J Acoust Soc 84:695–1708.

Ruggero MA, Rich NC, Robles L, Shivapuja BG (1990) Middle ear response in the chinchilla and its relationship to mechanics at the base of the cochlea. J Acoust Soc Am 87:1612–1629.

Saunders JC, Garfinkle TJ (1983) Peripheral anatomy and physiology I. In: Willot JF (ed) Auditory Psychobiology of the Mouse. Springfield, IL: Charles C Thomas, pp. 131–168.

Saunders JC, Summers RM (1982) Auditory structure and function in the mouse middle ear: An evaluation by SEM and capacitive probe. J Comp Physiol A 146: 517–525.

Schröter J, Poesselt C (1986) The use of acoustical test fixtures for the measurement of hearing protector attenuation. Part II: Modeling the external ear, simulating bone conduction and comparing test fixture and real-ear data. J Acoust Soc Am 80:505–527.

Schubert ED (1980) Hearing: Its Function and Dysfunction. New York: Springer-Verlag.

Schuknecht HF (1974) Pathology of the Ear. Cambridge, MA: Harvard University Press.

Searle CL, Braida LD, Cuddy DR, Davis MF (1975) Binaural pinna disparity: Another localization cue. J Acoust Soc Am 57:448–455.

Segall W (1943) The auditory region of the arctoid carnivores. Zool Ser Field Mus Nat Hist 29:33–59.

Segall W (1969) The auditory ossicles (malleus and incus) and their relationships to the tympanic: In marsupials. Acta Anat 73:176–191.

Segall W (1973) Characteristics of the ear, especially the middle ear, in fossorial mammals compared with those in the *Manidae*. Acta Anat 86:96–110.

Sellick PM, Patuzzi R, Johnstone BM (1982) Measurement of basilar membrane motion in the guinea pig using the Mössbauer technique. J Acoust Soc Am 72:131–141.

Shaw EAG (1974a) The external ear. In: Keidel WD, Neff WD (ed) Handbook of Sensory Physiology: Vol V/l: Auditory System. New York: Springer-Verlag, pp. 455–490.

Shaw EAG (1974b) Transformation of sound pressure level from the free field to the eardrum in the horizontal plane. J Acoust Soc Am 56:1848–1860.

Shaw EAG (1976) Diffuse field sensitivity of the external ear based on the reciprocity principle. J Acoust Soc Am 60:S102.

Shaw EAG (1979) Performance of the external ear as a sound collector. J Acoust Soc Am 65:S9.

Shaw EAG (1982) External ear response and sound localization. In: R Gatehouse (ed) Localization of Sound: Theory and Application. Groton, CT: Amphora Press, pp. 30–41.

Shaw EAG (1988) Diffuse field response, receiver impedance and the acoustical reciprocity principle. J Acoust Soc Am 84:2284–2287.

Shaw EAG, Stinson MR (1983) The human external and middle ear: Models and concepts. In: de Boer E, Viergever MA (eds) Mechanics of Hearing. Delft, The Netherlands: Delft University Press, pp. 3–10.

Shera C, Zwieg G (1991) Phenomenological characterization of eardrum transduction. J Acoust Soc Am 90:253–262.

Shrapnell HJ (1832) On the form and structure of the membrana tympani. London Med Gazette 10:120–124.

Siebert WM (1970) Simple model of the impedance matching properties of the external ear. Quarterly Progress Report of the Research Laboratory of Electronics, Massachusetts Institute of Technology, Cambridge pp. 236–242.

Siebert WM (1973) Hearing and the ear. In: Brown JHU (ed) Engineering Principles in Physiology, Volume 1. New York: Academic Press, pp. 139–184.

Silman S (1984) The Acoustic Reflex: Basic Principles and Clinical Applications. New York: Academic Press.

Sinyor A, Laszlo CA (1973) Acoustic behavior of the outer ear of the guinea pig and the influence of the middle ear. J Acoust Soc Am 54:916–921.

Sointseva GN (1973) Morphofunctional features of the outer ear of terrestrial, semi-aquatic and aquatic mammals. Reports of the 8th All Union Acoustic Conference, pp. 25–28 (in Russian).

Sointseva GN (1990) Formation of an adaptive structure of the peripheral part of the auditor (sic) analyzer in aquatic echo-locating mammals during ontogenesis. In: Thomas J, Kastelein R (eds) Sensory Abilities of Cetaceans. New York: Plenum Press, pp. 363–383.

Stephens CB (1972) Development of the middle and inner ear in the golden hamster (Mesocricetus auratus). Acta Otolaryngol Suppl 296:1–51.

Stinson MR (1985) The spatial distribution of sound pressure within scaled replicas of the human ear. J Acoust Soc Am 78:1596–1602.

Stinson MR (1986) Spatial distribution of sound pressure in the ear canal. In: Allen JB, Hall JL, Hubbard A, Neely ST, Tubis A (eds) Peripheral Auditory Mechanisms. New York: Springer-Verlag, pp. 13–20.

Stinson MR, Khanna SM (1989) Specification of the geometry of the human ear canal for the prediction of sound pressure level distribution. J Acoust Soc Am 85:2492–2503.

Teas DC, Nielsen DW (1975) Interaural attenuation versus frequency for guinea pig and chinchilla CM response. J Acoust Soc Am 58:1066–1072.

Teranishi R, Shaw EAG (1968) External ear acoustic models with simple geometry. J Acoust Soc Am 44:257–263.

Tonndorf J (1972) Bone conduction. In: Tobias JV (ed) Foundations of Auditory Theory, Volume II. New York: Academic Press, pp. 197–237.

Tonndorf J, Khanna SM (1967) Some properties of sound transmission in the middle and outer ears of cats. J Acoust Soc Am 41:513–521.

Tonndorf J, Khanna SM (1972) Tympanic membrane vibrations in human cadaver ears studied by time-averaged holography. J Acoust Soc Am 52:1221–1233.

Tonndorf J, Pastici H (1986) Middle ear sound transmission: A field of early interest to Merle Lawrence. Am J Otolaryngol 7:121–129.

Tonndorf J, Tabor JR (1962) Closure of the cochlear windows. Ann Otol Rhinol Laryngol 71:5–29.

Tröger J (1930) Die Schallaufnahme durch die äussere Ohr. Phys Zeits 31:26–47.

Unge M von, Bagger-Sjöbäck D, Borg E (1991) Mechanoacoustic properties of the tympanic membrane: A study on isolated Mongolian gerbil temporal bones. Am J Otolaryngol 12:407–419.

van der Klaauw CJ (1931) The auditory bulla in some fossil mammals: With a general introduction to this region of the skull. Bull Am Mus Nat Hist 62:1–352.

von Bismark G (1967) The sound pressure transformation function from free field to the eardrum of chinchilla. MS Thesis, Massachusetts Institute of Technology, Cambridge.

von Bismark G, Pfeiffer RR (1967) On the sound pressure transformation from free field to eardrum of chinchilla. J Acoust Soc Am 42:S156.

Vrettakos PA, Dear SP, Saunders JC (1988) Middle ear structure in the chinchilla: A quantitative study. Am J Otolaryngol 9:58–67.

Wada H, Kobayashi T (1990) Dynamical behavior of the middle ear: Theoretical study corresponding to measurement results obtained by a newly developed measuring apparatus. J Acoust Soc Am 87:237–245.

Waetzmann E von, Keibs L (1936) Theoretischer und experimenteller Vergleich von Hörschwellenmessungen. Akustische Zeitschrift 1:1–12.

Webster DB (1965) Ears of Dipodomys. Nat Hist 74:26–33.

Webster DB (1982) A function of the enlarged middle ear cavities of the kangaroo rat *Dipodomys*. Physiol Zool 35:248–255.

Webster DB, Webster M (1975) Auditory systems of Heteromyidae: Functional morphology and evolution of the middle ear. J Morphol 146:343–376.

Werner CF (1960) Das Gehörorgan der Wirlbeltiere und des Menschen. Leipzig, Germany: VG Thieme.

Wever EG, Lawrence M (1954) Physiological Acoustics. Princeton, NJ: Princeton University Press.

Wiener FM, Ross DA (1946) The pressure distribution in the auditory canal in a progressive sound field. J Acoust Soc Am 18:401–408.

Wiener FM, Pfeiffer RR, Backus ASN (1966) On the sound pressure transformation by the head and auditory meatus of the cat. Acta Otolaryngol 61:255–269.

Wilson JP, Bruns V (1983) Middle ear mechanics in the CF-bat *Rhinolophus ferrumequinum*. Hear Res 10:1–13.

Wilson JP, Johnstone JR (1975) Basilar membrane and middle ear vibration in guinea pig measured by capacitive probe. J Acoust Soc Am 57:705–723.

Woolf NK, Ryan AF (1984) The development of auditory function in the cochlea of the Mongolian gerbil. Hear Res 13:277–283.

Woolf NK, Ryan AF (1988) Contributions of the middle ear to the development of function in the cochlea. Hear Res 35:131–142.

Wullstein H (1956) The restoration of the function of the middle ear in chronic otitis media. Ann Otol Rhinol Laryngol 65:1020–1041.

Zuercher JC, Carlson EV, Killion MC (1988) Small acoustic tubes: New approximations including isothermal and viscous effects. J Acoust Soc Am 83:1653–1660.

Zwillenberg D, Konkle DF, Saunders JC (1981) Measures of middle ear admittance during experimentally induced changes in middle ear volume in the hamster. Otolaryngol Head Neck Surg 89:856–860.

Zwislocki J (1962) Analysis of the middle ear function. Part I. Input impedance. J Acoust Soc Am 34:1514–1523.

Zwislocki J (1963) Analysis of the middle ear function. Part II. Guinea pig ear. J Acoust Soc Am 35:1034–1040.

Zwislocki J (1965) Analysis of some auditory characteristics. In: Luce RD, Bush RR, Galanter E (eds) Handbook of Mathematical Psychology. New York: John Wiley and Sons, pp. 1–97.

Zwislocki J (1975) The role of the external and middle ear in sound transmission. In: Tower DB (ed) The Nervous System, Volume 3: Human Communication and Its Disorders. New York: Raven Press, pp. 45–55.

Index

This index combines both species and various topics. Individual species are referred to by common name. Scientific names are also indexed, but these give directions to common names used in the text.